# AGAINST THEIR WILL

# AGAINST THEIR WILL

## THE SECRET HISTORY
## OF MEDICAL EXPERIMENTATION
## ON CHILDREN IN COLD WAR AMERICA

Allen M. Hornblum,
Judith L. Newman,
and Gregory J. Dober

palgrave
macmillan

First published in hardcover in 2013 by PALGRAVE MACMILLAN® in the United States—a division of St. Martin's Press LLC, 175 Fifth Avenue, New York, NY 10010.

Where this book is distributed in the UK, Europe and the rest of the world, this is by Palgrave Macmillan, a division of Macmillan Publishers Limited, registered in England, company number 785998, of Houndmills, Basingstoke, Hampshire RG21 6XS.

Palgrave Macmillan is the global academic imprint of the above companies and has companies and representatives throughout the world.

Palgrave® and Macmillan® are registered trademarks in the United States, the United Kingdom, Europe and other countries.

ISBN: 978-1-137-27942-2 (paperback)

Library of Congress Cataloging-in-Publication Data

Hornblum, Allen M.
    Against their will : the secret history of medical experimentation on children in cold war America / Allen M. Hornblum, Judith L. Newman, Gregory J. Dober.
        pages   cm
    ISBN 978-0-230-34171-5 (hardback)
    1. Pediatrics—Research—Moral and ethical aspects—United States.
2. Children—Research—Moral and ethical aspects—United States. 3. Human experimentation in medicine—United States—History—20th century.
I. Newman, Judith L. II. Dober, Gregory J. III. Title.
RJ85.H67   2013
174.2'9892—dc23
                                                                    2012051032

A catalogue record of the book is available from the British Library.

Design by Letra Libre, Inc.

First PALGRAVE MACMILLAN paperback edition: September 2014

10  9  8  7  6  5  4  3  2  1

Printed in the United States of America.

*To A. Bernard Ackerman*

*Whose career exemplified the highest standards of ethics and medicine*

# CONTENTS

*Six pages of photographs appear between pages 150 and 151.*

# ACKNOWLEDGMENTS

**WE ARE INDEBTED TO THE MANY PEOPLE WHO ENABLED US** to chronicle this piece of medical history. In particular, we would like to thank the many former test subjects, relatives of test subjects, and physicians and medical researchers who endured our numerous meetings and phone conversations, shared their recollections with us, answered endless questions, and allowed us access to their privates papers, official documents, and photographs.

We would especially like to thank former "State Boys" Charlie Dyer, Gordon Shattuck, Austin LaRocque, and Joe Almaida for their recollections of Fernald's "Science Club" and tour of the institution. Also providing her time, her collection of documents, and her reflections on the Fernald story was Doe West, who chaired the Massachusetts task force that investigated the radiation experiments that occurred there. We also owe a debt of gratitude to Ted Chabasinski and Karen Alves for meeting with us and sharing their recollections about painful events that happened to them or to close family members.

Jessie Bly and Pat Clapp also deserve our thanks for taking the time to recall memories that became key aspects of the book. To doctors such as A. Bernard Ackerman, Hilary Koprowski, Constantine Maletskos, Cyril Wecht, Chester Southam, Avir Kagan, Jim Ketchum, Enock Callaway, and the many others who contributed to our understanding of medical research during the Cold War, we also owe a debt of gratitude.

We would like to thank the many librarians and archivists who assisted in our lengthy hunt for documents, journal articles, and the private papers of scores of physicians and medical researchers. To those helpful staffers at Harvard's Countway Medical Library, the American Philosophical Society, the

New York University Medical Archives, the University of Pennsylvania Medical Library, George Washington University Medical Archive, the Bancroft Library at the University of California (Berkeley), the Swarthmore University Archive, the College of Physicians of Philadelphia, the National Archives (College Park, Maryland), the Library of Congress, the Mandeville Library of the University of California (San Diego), the University of Pittsburgh Archives, and the Pennsylvania State Archive, we owe you a debt of gratitude.

Others deserving a note of appreciation for assisting us with procuring documents or better understanding the issues and events of this period include Shannon Fox, Alan Milstein, Joe Levin, Janet Albert-Herman, Eric Borseth, Jeff Kaye, Margot White, Paul Lombardo, Vera Sharav, Fred Misilo, Paul Lurz, Jack Power, William L. Rosenberg, and Joseph K. McLaughlin.

We would also like to thank the Penn State ACURA Research Program and students Sharayah Wilt, Rada Khurgun, Sheba Favardin, and Karishma Minocha for their support and assistance in gathering relevant documents. We also appreciate the research stipends from the Rubin Fund and a Penn State Development Grant.

Another individual deserving our thanks is the late Sidney Newman who consistently throughout his long life saw children with mental challenges as most deserving of respect and care. He taught integrity by example.

And lastly we would like to thank our agent Jill Marsal and our editor at Palgrave, Luba Ostashevsky, for recognizing the importance of this work and for their help in guiding us to the publication finish line.

# INTRODUCTION

## *"They'd Come for You at Night"*

*It is better for all the world, if instead of waiting to execute degenerate offspring for crime, or to let them starve for their imbecility, society can prevent those who are manifestly unfit from continuing their kind.*

—Oliver Wendell Holmes

**"THEY TOLD ME I SHOULDN'T HAVE KIDS AND I SHOULDN'T** get married. They said I might have a defect, that I had something wrong with me. They said, 'You aren't stabilized and you shouldn't have kids because of what you have.' I didn't know what I had. Nobody ever told me. I thought I was like everyone else. But I was only a kid, I had no interest in getting married; I was only fourteen years old. And to tell you the truth, I just wanted out of there," recalls Charles Dyer of his unhappy days growing up in a series of bleak Massachusetts institutions.[1] It was 1954, and America was in the thick of the Cold War. Communism was spreading around the world. The Korean War had just ended; the Army–McCarthy hearings dominated political discussions in Washington.

It was also a significant time for young Charlie. He had just completed his first year in a large, forbidding state institution that was once known as the Massachusetts School for Idiotic and Feeble-Minded Youth. Over one hundred years after its founding as a refuge for high-functioning disabled children, it was still home to what many people at the time considered a motley collection of societal refuse—"defective" humans who were commonly referred to as "morons," "mongoloids," and "gargoyles," not to mention your everyday "idiots" and "lunatics." By today's standards, however, some, like Charlie, would seem little different from the rest of us.

Charlie was born in 1940 in Auburn, a small town just south of Worcester, Massachusetts. His mother was a state employee and his father a long-haul trucker and lumberjack who spent much of his time working in Canada. Both were alcoholics whose parenting skills left much to be desired. "I was one of eight kids," says Charlie, recollecting his troubled youth. "There were five boys and three girls. My father was kind to us, but he was rarely home. We were lucky to see him once every six months. My mother, on the other hand, was a nasty drunk. She was particularly mean toward me and picked on me a lot. She often slapped me in the face, and a few times she nearly pulled my arms out of their sockets."

Life outside the Dyer home wasn't much more appealing. Charlie never progressed past the first grade. Following some disciplinary problems and a third unsuccessful attempt to master reading, writing, and arithmetic, a state social worker dropped the eight-year-old off at the Lymon School for Boys in Westborough, the nation's oldest reform school. The experience was traumatic for the slight, blue-eyed blond boy.

"They used to really come down on you when you got out of line," says Charlie of Lymon's hard-nosed staff. "It wasn't a place to help kids, that's for sure. Kids got punished for everything; most of the time by having staff members sit on you until you couldn't breathe. They were rough with the kids. And that's not all; they used to take all the best stuff that was donated to the place and keep it for themselves. The best clothes, the best athletic gear; we were left with the worst of everything."

Charlie was eager to learn and initially wondered what school would be like at Lymon. "There was really no schooling there," he says, "but they did make us work. I was only a little kid, but they hooked a harness around me like you do a horse and made me and other kids pull a plank with a rug wrapped around it to wax the wooden floors. Back and forth we'd go for a couple hours until the floor shined. And if you didn't do it right or got out of line, they'd whack you on the ass with a wooden paddle. Each paddle had a child's name stamped on it, and they'd grab one if they thought you deserved it."

Lymon was followed by a succession of other institutional schools including the Home for Little Wanderers in Brookline, Massachusetts. Some, such as the Metropolitan State Hospital, had bars on the windows, which deeply impressed young boys like Charlie.

By the time he was in his early teens, Charlie was shifted to the Walter E. Fernald State School in Waltham, an institution with a long history steeped in the principles of eugenics. The school was founded by Samuel Gridley Howe, an antebellum doctor and reformer. He operated under the belief that disadvantaged and disabled children could be educated to lead relatively productive, independent lives if taught the right skills and proper decorum. From 1848 when it was founded in South Boston to its move to bucolic Waltham, the school continued to grow in both size and stature.

By the last couple of decades of the nineteenth century, however, Fernald—and scores of other schools like it throughout the nation—were being buffeted by the strong winds of philosophical change. Instead of seeing their role as training the disadvantaged, the reformers wanted to solve the problem. They demonized the mentally and physically challenged as a drain on society and deemed them unsuitable for reproduction. Those who were severely retarded and physically handicapped were the first targets, along with people who were retiring or shy or who stuttered, and many non-English-speaking individuals, regardless of skill or intelligence. "Feeblemindedness," as one critic of the movement has written, "was truly in the eye of the beholder and frequently depended upon the dimness or brightness of a particular moment."[2] Those now thought "genetically unfit" would be sequestered in ever larger and more austere institutions by those who had no real intention of releasing them back into society. They would be warehoused indefinitely, and some would face additional horrors.

Initially, this unforgiving and pessimistic movement had a positive, constructive bent. Given the name "eugenics" in the early 1880s by Sir Francis Galton, a cousin of Charles Darwin, the nascent campaign married the latter's theories of natural selection to Gregor Mendel's revolutionary work in genetics. A sprinkling of Herbert Spencer's "survival of the fittest" thinking was thrown in for additional "naturalist" seasoning. The upshot was a movement built on a simple and seemingly incontrovertible scientific principle that all human characteristics, from intelligence and morality to character and longevity, were grounded in one's biological makeup. Once the die was cast, it was generally thought, little could be done to alter it.

Galton believed that there were ways to capitalize on the very best society had to offer. Very simply, if one was part of the elite—those of heroic stature and a talented disposition—it was one's moral duty to have "a larger number of children than the average family." Men like Napoleon, Beethoven, Van Gogh, Pasteur, and Bismarck shaped history and moved mankind forward, according to Galton, and therefore should be encouraged to procreate. The best individuals should be the finest and most prolific breeders. The result would be the advance of civilization and concomitant progress in the arts, science, literature, and politics.

There were those in the movement, however, who were less focused on this heroic class than on its opposite—the human wreckage or flotsam and jetsam of society who were a constant drag on civilization and a persistent impediment to reaching its potential. Charles Benedict Davenport, one of the movement's more ardent proselytizers, regarded those with what he considered weak traits, deep-seated character flaws, and severe physical impairments—he often referred to them as "the submerged tenth" and "defective germ plasm"—as destined to become beggars, lunatics, thieves, and prostitutes who would disrupt society and eventually fill the nation's prisons, poorhouses, and hospitals.[3] In time, impoverished urban dwellers and dispossessed farmers from Maine to California, Eastern European immigrants, southern blacks, Russian Jews, illiterate Italians, desperate Mexicans, prostitutes, pickpockets, alcoholics, the mentally compromised, and anyone else who did not resemble the Northern European archetype would find themselves potential victims of eugenic fanaticism.

By World War I, the reform movement had come to focus on what it called "degeneracy" and the continuing decline of the human stock. The goal of negative eugenics, as it would eventually be known, would be to prevent "the reproduction of inferior stock."[4]

For those at the bottom of the social ladder, such as individuals confined at Fernald and similar schools, this news was particularly cruel. They were viewed as the most glaring examples of genetic failure and were destined to live compromised lives, work in menial labor, and rely on the generosity of strangers—or, increasingly, the state. It would not be long, however, before the "mentally deficient," especially those considered "morons, imbeciles, and idiots," were no longer merely a social burden but were also seen as a social menace and as a growing percentage of the nation's paupers, prostitutes, thieves, and criminals.[5]

It was also believed that society would do well to isolate such individuals, limit their exposure to temptation, and implement policies and laws that would prevent their procreation and their unhealthy, antisocial tendencies. Thousands were sterilized and thousands more institutionalized, to spend their days locked away in terminal boredom or, in later years, a drugged stupor. The lucky ones in this dystopian environment learned basic skills, such as sewing, cleaning, and gardening. Institutions—many designed for children—such as those at Vineland (New Jersey), Letchworth Village (New York), and Pennhurst (Pennsylvania) would take their place in the pantheon of human horror shows containing every type of mental defect and physical deformity imaginable.

The eugenic fervor in the United States would gradually wane during the 1920s and dissipate further during the Great Depression while speeding up and taking a more malevolent form in Nazi Germany. At the same time, there were still orphanages and institutions for the disabled, and the treatment given to inmates there still showed the influence of eugenics. Although no longer a philosophical force, the eugenics movement had done its job: thousands of Americans had been dehumanized and thoroughly devalued.

An argument can be made that conditions and so-called therapeutic treatments grew considerably more punitive and frightening over the decades into the 1940s, the postwar years, and throughout the Cold War. Surgical sterilization for both men and women was widespread, practiced routinely in more than half the states and endorsed by the highest court in the land. Shock treatments, including insulin, Metrazol, and electroshock, were now institutional staples, and lobotomy had become the new therapeutic rage among neurologists, psychiatrists, and frustrated, overworked ward physicians who were desperate to cure the thousands of mentally maimed wasting away in hospital and asylum wards throughout the country. Many doctors were willing to try anything that would turn off the faucet that was filling the nation's institutions.

World War II would have a profound effect on scientific research in America. And those abandoned by their families and warehoused in large state institutions would discover that they were finally of interest to someone—medical investigators searching for a wide assortment of preventives and treatments. Saturated in several decades of dehumanizing eugenics propaganda, enterprising researchers aggressively sought out facilities that provided access to their residents. A convenient pattern and ethos was established—an unquestioning open-door policy—that would only become more accepted and ingrained in the culture of the research arena with the passage of time.

By the time young Charlie Dyer entered Fernald in the early 1950s, eugenics may have been a fading memory more closely associated with its Third Reich excesses, but remnants were still visible in this country.[6] It was quickly being replaced, however, by another even more ominous ideological challenge. This new threat was neither homegrown nor genetically based but foreign in origin and, most important, designed to "bury" the American capitalist system.[7] Long suspected and guardedly watched during the first half of the century, Communism in the postwar years was seen as a rapidly growing menace and serious challenge to world stability. Stalin and the Soviet army had withstood the Nazi invasion, developed an atomic bomb years before anyone predicted, and were secretly preparing to launch the world's first artificial satellite into space. The Soviet Union was also making no secret of its ideological and territorial expansion. Events in Eastern Europe and China provided further proof of that. The world's largest nation, Nationalist China, had been "lost" the same year the Soviets had developed the bomb, and "Red China" was now aggressively following its Marxist partner in fostering a Communist future for millions around the world.

As one close observer has written of the period, two Communist giants were now allied. The Soviet Union, already a menacing threat to the West, now had an even larger nation on its side. "The contest between Communism and the free world had taken an ominous turn in favor of the former."[8]

Alarmed by a new, threatening challenge, American governmental, cultural, and scientific institutions reacted swiftly and in a multitude of ways. Children like Charlie Dyer often found themselves in the crosshairs of a ramped-up medical and scientific crusade designed to conquer our Marxist-Leninist competitors as well as age-old public health concerns such as polio, tuberculosis, and cancer.

For the thousands of institutionalized children, the competing economic systems, vitriolic Cold War rhetoric, and pharmaceutical expansion were the furthest things from their minds. They had more immediate issues to contend with including a stark, unforgiving environment, harsh overseers, and the threat of sexual and physical violence.

After some of the previous institutions he had been in, Charlie Dyer had a fleeting sense of optimism about Fernald. "When they brought me there," he says, "I thought it would be better. There were no bars on the windows, and I hoped they'd let us out to play once in a while. Maybe they'd have a real school, and I'd finally learn to read and they'd teach me a trade so I could get a job one day." Unfortunately, Charlie's dreams were quickly dispelled.

"They never taught me anything there," Charlie explains. "We wanted to learn about things, but the teachers just sat on their ass. They gave us coloring books and never taught us anything. We'd get homework to do but [they] wouldn't show us how to do it. They'd say, 'Figure it out for yourself.' They were no help at all. They gave us medicine to calm us down, but that just put us to sleep when we should have been learning something. I got so angry I told them, 'The hell with you. I'll learn it myself.'"[9] In fact, most of the boys in Fernald never learned to read.

What he and the others did learn was how to stuff and sew mattresses, knit blankets, and make brooms. In summer, they were marched out to the surrounding fields and picked fruit and vegetables all day, like subsistence sharecroppers. But they weren't sharecroppers; they were children, children with a depressing past and an equally dismal future. Their grim living conditions were made all the worse by periodic attacks—some sexual in nature—by older boys at the facility. Those paled in comparison to the frequency—and sheer terror—of the assaults perpetrated on them by Fernald staff.

"They'd come for you at night when everyone was sleeping," said Charlie. "I was small; most of the kids were bigger. But they'd pull a kid out of bed and take him to another room and make him do things. It happened to all of us. That's the type of people they hired there."

One day, however, Charlie and a number of other boys were called down to a meeting with some well-dressed, professional-looking outsiders who began promising them trips to Fenway Park, occasional excursions to the seashore, and a tour of the laboratories at the Massachusetts Institute of Technology. "They told us if we joined this new club," says Charlie, "they'd give us a Mickey Mouse watch and send us on field trips and to baseball games. We jumped at the chance to get in the Science Club. We would have done anything to get out of Fernald, even if it was just for the day." For a kid who had been shipped around a lot and called mentally "retarded" most of his life, this was a very welcome change of pace. All the boys felt that way.

Those who volunteered for the new club were immediately separated from the others and sequestered on a ward where they were monitored daily, forced to give blood on a regular basis, and ordered to urinate and defecate in glass jars and endure an assortment of metabolic tests and physical examinations. The panoply of tests and strange procedures quickly cooled the ardor of even the most enthusiastic club participants. "I hated the needles," recalls Charlie, "and I wasn't allowed to see my friends anymore. They said we were just testing vitamins, but it made no sense to me. And the few trips we went on weren't worth the pain and the isolation."

Charlie told the doctors he wanted out; he no longer wanted to be in the Science Club. But they told him he couldn't leave; no one could. All the boys were obligated to remain involved until the doctors told them it was okay to leave the program. After repeated attempts to gain his freedom, Charlie became so desperate he did something that nearly cost him his life. His spur-of-the-moment actions sparked a panic at the school and created an incident that eventually became part of Fernald lore.

Not until decades later, however, would Charlie and the other Science Club members—not to mention the rest of the nation—learn the truth: the club was actually a front that enabled private and public sector entities, including universities, corporations, and the government, to use the Fernald children in a series of secret radiation experiments. For nearly twenty years—from the 1940s, when the boys' oatmeal breakfasts were contaminated with radioisotope-laced milk, to the 1960s, when more serious Cold War experiments were orchestrated—the children were used like caged guinea pigs in a laboratory. The Fernald story, as well as other revelations of unethical medical research at the time, would shock the collective conscience of America and spark a major government investigation into clandestine radiation studies in the aftermath of Hiroshima and Nagasaki. It would also cement, among former Science Club members as well as their friends and families, a lasting distrust of doctors and the medical establishment.

Many test subjects at the dawn of the Atomic Age and throughout the decades that followed, as the public would come to learn, were children. Some were only days old; some were cognitively and physically impaired. We believe this tragic chapter in twentieth-century medical history—the purposeful and systematic exploitation and commodification of society's weakest members—is worthy of exploration and comment.

**THE SAD HISTORY OF CHILDREN**, especially institutionalized ones, being used as cheap and available test subjects—the raw material for experimentation—started long before the Atomic Age and went well beyond exposure to radioactive isotopes. Experimental vaccines for hepatitis, measles, polio, and other diseases; exploratory therapeutic procedures such as electroshock and lobotomy; and untested pharmaceuticals such as curare and Thorazine were all tested on children in hospitals, orphanages, and mental asylums as if they were some widely accepted intermediary step between chimpanzees and humans. Occasionally children supplanted the chimps.

Bereft of legal status or protectors, institutionalized children were often the test subjects of choice for medical researchers hoping to discover a new vaccine, prove a new theory, or publish an article in a respected medical journal. Many took advantage of the opportunity. One would be hard pressed to identify a researcher whose professional career was cut short because he incorporated week-old infants, ward-bound juveniles with epilepsy, or those with profound retardation in his experiments. Involuntary, nontherapeutic, and dangerous experiments on children were far from unusual or dishonorable endeavors during the twentieth century. The practice was widely accepted, rarely questioned, and integral to the phenomenal growth of medical research and human experimentation during World War II and the Cold War that followed.

Institutionalized children, like other vulnerable populations including prisoners, soldiers, hospital patients, and those with mental illness, were an attractive wellspring of opportunity for enterprising doctors and scientists. Few in the medical profession thought such practices unseemly or problematic; even fewer spoke up to express opposition or voice moral outrage at these widely known and frequently cited exercises. In fact, doctors aggressively pursued relationships with superintendents, headmasters, and matrons of institutions that held vulnerable populations. Doctors quickly discovered that access to institutionalized populations could springboard them to lucrative contracts with drug companies and great wealth. The financial incentives became so enticing that some physicians gave up their private practices to conduct large-scale clinical trials full time.[10]

Only in recent decades, however, have we begun to shine a light into the dark closet of medical ignomiy and illuminate the many stories of vulnerable subjects who were chosen for their perceived inadequacies and handicaps.

Regrettably, for researchers seeking suitable test sites, powerlessness was every bit as enticing a feature as regimentation and convenience. Bluebloods, members of the *Social Register,* and those attending elite prep schools were rarely included in clinical trials. Those at the bottom of the socioeconomic ladder, however, weren't nearly so fortunate. Social, racial, physical, and intellectual disabilities stamped such individuals as "throwaway people," classes of flawed or "defective" individuals without value. Except, of course, as grist for the research mill—"material" in the medical journal lexicon of the day. Ironically, they had been selected by and would suffer at the hands of some of America's best and brightest scientific minds.

Few at the time seemed to have a problem with it. Not until the mid-1960s was there an inkling of doubt about the exploitative practices. As one eager investigator of the era who regularly incorporated retarded children, prisoners, and indigent senior citizens in his clinical trials nostalgically commented, "No one asked me what I was doing. It was a wonderful time."[11]

Through the efforts of scholars, investigative journalists, and medical ethicists, critical pieces of history have been reclaimed, which we hope will ensure that such grievous incidents of unethical medicine never again occur. *Bad Blood,* James Jones's 1978 account of several hundred unschooled Alabama sharecroppers going untreated in a four-decade-long syphilis study by the US Public Health Service, was the first examination of vulnerable populations being deceived, mistreated, and medically exploited.[12]

There have been other books about medical experimentation on institutionalized children, but few have attempted a pointed and frank conversation about the motivating factors that allowed so many physicians and researchers to subscribe to a system and professional ethic that routinely placed infants and children in harm's way. Or why society seemed indifferent to the practice, especially when it was occurring in the one nation that felt it necessary to punish the Nazi doctors for their many ethical transgressions and brutal medical experiments. The ethically toxic alchemy of the eugenics movement, World War II, and the threat of Communism during the Cold War played a significant part in allowing some of our best and brightest to cavalierly exploit some of our youngest and those most deserving of protection. An accurate accounting and discussion of this sad phenomenon is long overdue.

We hope this volume is viewed as a positive step in that important undertaking. It is also our goal that the work provide a voice—albeit a belated one—for the thousands of children who were conscripted to serve in the name of science. The crucible of scientific advancement may have brought forth increased knowledge during the twentieth century, but the young, docile, mute soldiers in the campaign witnessed more burden than acclaim. As human guinea pigs, they received no notice, reward, or appreciation. Their story is not a pleasant one, but it is one that deserves to be told.

# ONE

# THE AGE OF HEROIC MEDICINE

*"At Their Best, Medical Men Are the Highest Type Yet Reached by Mankind"*

THERE ARE FEW TODAY WHO WOULD NOT IMMEDIATELY REC-
ognize the names Babe Ruth, Jack Dempsey, and Red Grange. Their iconic status as sporting greats from what is usually referred to as the Golden Age of Sports is arguably unrivaled, a testament to their physical accomplishments and their stature as Mount Olympus–caliber athletic competitors as well as the sports-obsessed era in which they competed. Even a horse, Man o' War, managed to cement his name and stirring victories into the consciousness of adoring sports fans during the 1920s.

Books, newspaper and magazine articles, and movie theater newsreels repeatedly showcased the physical talents and athletic exploits of the era's much-celebrated champions, so it was no surprise that comic books would devote numerous titles and issues to their accomplishments on the gridiron, diamond, and track.

Surprisingly, however, in addition to splendid athletes, courageous crime detectives, and all-powerful science fiction characters, some of the world's leading medical and scientific minds would grace the titles and covers of children's magazines. They would not only add some cultural and substantive heft to the lowbrow but extremely popular newsstand magazines but also under-score the rising status of physicians and highlight the importance of the medi-cal profession in contemporary society. By World War II, for example, gallant generals and daring soldiers were cover-story topics, but brilliant medical sleuths on the cutting edge of new treatments and vaccines held their own and often shared newsstand space with them.

American medicine had come a long way. As Harvard sociologist Paul Starr has written, prior to the twentieth century the role of doctor did not con-fer a clear and distinct class position in American society.[1] In the nineteenth century, most medical men were often no more than autodidacts, veterans of an apprenticeship system that required no formal education. The result was a profession that had little status and modest earning potential.

By the early years of the twentieth century, however, the prestige and income potential of the medical profession had witnessed enormous gains; in fact, medicine had become a highly desirable career choice. In 1925 physicians ranked just behind bankers and college professors and just ahead of clergymen and lawyers according to a survey of high school students and teachers. By the 1930s medicine had jumped to the top of every occupational category and would remain there. Only one position exceeded physicians: "Justice of the U.S. Supreme Court."[2]

Sobering accounts of medical sacrifice and the excitement of triumphant discoveries, as well as an appreciation of medical history, took off in the 1920s. Profiles of the men and women who played a role in that history became a key part of the narrative. One of the first and arguably most influential of the medical success story chroniclers during this period was Paul de Kruif. Though little remembered today, de Kruif was instrumental in illuminating the challenges and sacrifices made by the great men and women of science who had committed their lives to fighting disease and pestilence.

After serving in World War I and participating in the hunt for Pancho Villa in Mexico, de Kruif earned a doctorate in bacteriology at the University of Michigan and became a research fellow at New York's prestigious Rock-efeller Institute, a well-recognized powerhouse of scientific research and tal-ent. De Kruif also nurtured a side interest; he wanted to become a writer. He began to chronicle what he was observing in the lab and writing illuminat-ing vignettes about the men who were searching for the answers to age-old mysteries. De Kruif began with a series of anonymous articles in *Century* magazine in 1922. The articles underscored his conflicted feelings about the profession, both the hero worship and his increasing disenchantment. At Rockefeller, de Kruif rubbed shoulders with the elite of contemporary medical research. Many were Nobel Prize winners or soon would be. But despite the great minds and many awards associated with the Rockefeller Institute, de Kruif was overcome with a growing skepticism. As he would recount decades later, "The years wore on but the hoped-for parade of cures did not come off. Could it be," he asked, "that the slot machine had turned out to be a one-armed bandit . . . ?"[3]

Weighed down by doubts and personal issues that included a failing marriage and regrets on leaving a child behind, de Kruif took a leave of absence from his assignments at Rockefeller in 1922 and began to write about medical research as he knew it and the great men and institutions that were involved in the battle against disease. Not all of it would be positive. One 1922 magazine project, initially titled "Doctors and Drugmongers," and his first book, *Our Medicine Men*, illuminated the pretentiousness and naïveté of American medicine. The works drew public interest, at least one lawsuit, and his prompt dismissal from the Rockefeller Institute. He was fired "for irreverently daring to write—a la Henry Mencken—spoofing Rockefeller science and for disrespect to medicine's holy of holies."[4]

De Kruif considered his firing a "self-inflicted kick in the teeth" that spurred his move from science to journalism. Before tackling another subject on his own, he would perform the duties of a scientific aide-de-camp and idea repository for one of the literary giants of the period, the novelist Sinclair Lewis. The much-admired literary storyteller was searching for his next project when he met de Kruif. The relationship between the master novelist and the aspiring journalist would prove critical to one of Lewis's greatest literary triumphs. As Lewis listened to de Kruif's stories of the institute, he increasingly perceived the seeds of a storyline, one that had all the makings of a landmark work. De Kruif agreed. In the hands of a master storyteller like Lewis, de Kruif's account of scientific research at the highest levels could prove an "epic of medical debunkology."[5]

The result was the American classic *Arrowsmith*, a Pulitzer Prize–winning novel whose heroic leading man was a dedicated research scientist, supposedly "the first of consequence in American literature." According to medical historian Charles Rosenberg, Martin Arrowsmith was a "new kind of hero, one appropriate to twentieth century America."[6] The book's unusual theme and hero not only appealed to aspiring science buffs but, surprisingly, also resonated with the American public.

De Kruif and Lewis had created the protagonist Martin Arrowsmith, a genuine scientist with a religious-like fervor for the truth. His tightly conceived and meticulously coordinated experiments would be untainted by professional aspirations, commercial distractions, or government restrictions. Contemporary American life—particularly as it was evolving in the Roaring Twenties—was pervasive in its demeaning and degrading materialism and its endless status seeking; even pure science had been infected and tarnished. Lewis and de Kruif had their character rebel against such corrupting influences and seductions; they ultimately had Arrowsmith jettison his family and material considerations for a simpler life that was consistent with pure scientific research.[7]

*Arrowsmith* was a great success, won its author a Pulitzer and even greater notoriety, and was made into a successful Hollywood film.[8] It would also give most Americans their first peek into the culture of American medicine.

With the Lewis project completed, de Kruif was in search of a project of his own.

"One day," as he would later write, he began thinking about "Leeuwenhoek, the first of the microbe men."[9] Soon he had completed a manuscript containing the contributions of twelve great scientists, from Antonie van Leeuwenhoek, who first peered into a fantastic new world of microscopic organisms in the seventeenth century, to Paul Ehrlich, who discovered his groundbreaking 606 medicinal recipe that effectively destroyed spirochaetes bacteria, or the cause of syphilis, three centuries later. They were microbe hunters in de Kruif's book of that name, brilliant, dedicated men of science who had conquered insidious diseases and saved the lives of untold millions. In recounting their difficult quests, their self-sacrifice and intellectual challenges, and finally the grandeur of their discoveries, de Kruif transformed a piece of staid laboratory history into a series of true adventure sagas that appealed to readers who normally had little interest in science.

Harcourt Brace—the book's publisher—initially was not optimistic about de Kruif's "off-beat opus"; medical research tomes weren't big sellers. The company thought it would be "lucky if it sold-out its first printing," a modest 2,800 copies. As de Kruif happily recalled years later, "Immediately upon publication the sales of the book exploded in our faces." Favorable reviews from prominent literary and social critics propelled it along. "One of the noblest chapters in the history of mankind," roared Henry Mencken. "A book for those who love high adventure, who love clear, brave writing," wrote William Allen White. The book shot up the nonfiction bestseller lists during the summer of 1926, quickly passed 100,000 copies in sales, and "became one of the big nonfiction books of the decade."[10] De Kruif had written the right book at the right time for a hero-worshipping age.

Much of its success, no doubt, was de Kruif's accessible, breezy style of historical portraiture. By combining biographies of serious-minded men who had labored long and hard to break new scientific ground and a writing style that was more typical of tales of the Old West, de Kruif discovered the secret to seducing young and old readers alike.

De Kruif was not just an academic who could craftily turn a phrase and tell an interesting story; he recognized nuance, grappled with moral boundaries, and valued visionary leadership, even when those attributes were sometimes in conflict. He appreciated Walter Reed's creative and intrepid pursuit of the fearsome mosquito that spread death and disease throughout the tropics, but he also recognized that the great American physician was rolling the dice with other people's lives—in some cases the lives of friends and colleagues. As de Kruif wrote of Reed and his deadly experiments, "To make any kind of experiment to prove mosquitoes carry yellow fever you must have experimental animals, and that meant nothing more nor less than human animals." De Kruif did not always sugarcoat unpleasant facts; he let readers know that Reed was involved in human experimentation and that such research often ended tragically. There was no getting around it, de Kruif informed his readers; that was the nature of human research. It was "an immoral business."[11]

The thousands of Americans who read *Microbe Hunters* in the 1920s and the decades that followed were told of the trade-offs involved with such historic undertakings. Solving an age-old riddle that had snuffed out millions of lives wasn't child's play; experimenting with tropical diseases was a serious and deadly affair. Reed, according to de Kruif's account, understood that challenge—he had "not one particle of doubt he had to risk human lives"[12] to solve the mystery of the deadly yellow fever. "'You must kill men to save them!'" argued the great microbe hunter. "Never was there a good man," wrote de Kruif admiringly, "who thought of more hellish and dastardly tests."[13]

But de Kruif went out of his way to assure readers that men like Dr. Reed and his ilk were above suspicion, guiltless. Yes, giving test subjects yellow fever would be considered "murder" by some, but Reed's zealousness and his willingness to put people at risk were understandable. It was crucial that medical science determine, by any means necessary, if mosquitoes caused yellow fever.

The book caused many Americans to confront the unattractive calculus of medical research. It wasn't pleasant, but it paid dividends—thousands of innocent lives eventually would be spared. De Kruif's heroic accounts of great men doing dangerous things would also illuminate the majesty as well as the desperation of the various test subjects, the experimental guinea pigs. With a strong whiff of American chauvinism, de Kruif praised those educated Westerners who understood the risk, appreciated the historic moment, and offered themselves as test subjects. They were accorded heroic status. "'We volunteer solely for the cause of humanity and in the interest of science,'" says a brave army private, according to de Kruif, during one solemn scene. "'Gentlemen, I salute you,'" replies an obviously appreciative Major Reed. In the author's eyes, all such military volunteers were "first-class, unquestionable guinea-pigs, above suspicion and beyond reproach."[14]

But many more would be needed to test the theory and solve the deadly riddle. The "dastardly experiment" designed by the "insanely scientific Walter Reed"[15] required ever more subjects. If there weren't enough "Americans who were ready to throw away their lives in the interest of science," then "there were ignorant people," those just arriving in "Cuba from Spain and who could very well use two hundred dollars."[16] In describing the induction of these "mercenary fellows" as test subjects, although they would suffer the same raging fevers, agonizing discomfort, and fits of vomiting and diarrhea, they never came off as quite as gallant as the American soldiers and doctors who volunteered for the research program.

De Kruif's triumphant account of the landmark yellow fever experiments and the driven—almost heartless—attitude of the chief medical investigator resonated with readers just as Martin Arrowsmith's fictional quest for scientific purity had captured the hearts and minds of readers a year earlier.

The message of *Microbe Hunters* was clear: Great men like Pasteur, Reed, Theobald Smith, and Paul Ehrlich were a rare breed. But for all their skill, training, and dogged pursuit of that deadly microbe or magical elixir, their mission was infinitely complex, the challenges multifaceted, and the trail of disease and death a daily occurrence. They had to make difficult decisions, decisions that might look cavalier and callous to the casual observer. And yes, on occasion they'd even appear cruel and "less human" in the eyes of some. But if society was to benefit in the long run, these extraordinary men of medicine—arguably the best and brightest of their class—must have the freedom to operate and to proceed unencumbered by moral tastes of the moment or by bureaucratic constraints. Granted, some unfortunate souls would lose their lives in these roulette-like experiments, but then, as de Kruif argued, "the microbe hunters of the great line have always been gamblers." He insisted that we not dwell on the negative but think of the heroic mission—focus on "the good brave adventurer and the thousands he has saved."[17]

And many did just that.

IN ADDITION TO THE DOUBLE SHOT of medical research triumphalism offered by Lewis and de Kruif, 1926 would witness a literary trifecta as Harvey Cushing's biography of William Osler, the most admired physician in English and American medicine, would win the Pulitzer Prize. Rarely had literary and historical works about the culture of medicine and the men behind the healing arts so captivated readers. The staid professions of medicine and medical research had received a shot in the arm, if not a dramatic makeover. Those in the business of the "wholesome adventure" genre took notice and subsequently propelled popular heroic narratives to include men in white lab coats alongside athletic champions, battlefield heroes, and comic book characters.

Comic books, however, were only one of many entertainment and informational sources that sought to capture the drama of dedicated doctors combating deadly diseases and laboratory researchers vanquishing public health maladies in the hope of discovering miracle cures and saving lives. As early as the 1920s, "medical history images, and stories came to be widely disseminated in popular books and magazines, commemorations, Hollywood films, children's literature, radio dramas, schoolbooks, corporate advertising, and the then-brand-new genre of comic books."[18]

The phalanx of media sources trumpeting the devoted and occasionally death-defying actions of medical researchers heightened the image of the medical profession not only with better-educated Americans but with general readers and their children. Moreover, it also helped solidify those middle decades of the twentieth century as the Golden Age of American Medicine and foster a greater public appreciation for doctors, their profession, and their soaring status in society. For doctors to be lionized with sports icons like Lou Gehrig and war heroes like General George S. Patton was no small achievement.

Louis Pasteur and Theobald Smith may not have been able to turn out tens of thousands of spectators in athletic coliseums the way Jim Thorpe, Jimmie Foxx, and Gene Tunney did or have the avid following of comic book superheroes, but they held their own in the public-esteem category. More important, they were resolute in their commitment to doing good work, and their efforts were proving beneficial for mankind. There were definite reasons to celebrate their lives and their contributions. A halo effect had gradually settled upon the research community's best and brightest stars.

Book readers and radio listeners were increasingly learning about Louis Pasteur and Edward Jenner as well as scientists they had never heard of, such as Elie Metchnikoff, Waldemar Haffkine, Horace Wells, Émile Roux, Giovanni Battista Grassi, Ronald Ross, and Robert R. Williams—men who had made significant contributions in fighting a host of dreaded diseases. The contributions of women were also celebrated. Florence Nightingale, Clara Barton, Elizabeth Blackwell, and lesser-known figures such as Jeanne Mance and Mary Walker were for the first time receiving significant attention. Even a Siberian husky named Balto was feted for leading a team of sled dogs carrying medicine to Nome, Alaska, where a diphtheria epidemic had broken out.

With the rewards so much greater than the penalties and with medical ethics in its infancy, researchers proceeded in a very laissez-faire manner. Experimentation on humans during the early decades of the twentieth century was widely accepted with relatively few speaking in opposition. Misuse and miscalculation in the handling and treatment of test subjects were considered more of an oversight or a minor error in judgment than an egregious criminal act. That was especially the case if the research resulted in a therapeutic breakthrough or a valuable piece of new information. It should be understood that doctors did not want to damage their patients—as a profession they were sworn to do no harm—but if they committed dastardly acts, they were more easily pardoned if something positive had come of the exercise. Experiments on humans were usually excused if the results of the study were substantial, the process had an element of science to it, and the physicians were correct in their expectations.[19]

The results were increasing instances of questionable experiments being performed by members of the medical profession. In addition to numerous experimental studies on men and women and on more vulnerable populations such as hospital patients, asylum residents, and prisoners, children were frequently used as research material. In 1895, for example, Dr. Henry Heiman, a New York pediatrician, purposefully infected a four-year-old "idiot with chronic epilepsy," a sixteen-year-old "idiot," and a twenty-six-year-old man with advanced tuberculosis with active gonorrhea germs.[20] That same year, while exploring the effectiveness of vaccines against smallpox, Walter Reed and George M. Sternberg used children at several New York City orphanages for their studies.[21] It was not uncommon at the time for doctors to evaluate the immunity claims of a smallpox vaccine by injecting children with the active virus after vaccinating them. Even infants were used in research studies; for example, two-day-old babies were fed bismuth, a metal-like substance used in the manufacture of some pharmaceuticals, and then exposed to extensive X-rays to chart the course of different foods in their stomachs.[22] And in 1907, three associates of the William Pepper Clinical Laboratory at the University of Pennsylvania used well over a hundred children under the age of eight at the St. Vincent's Home for Orphans, a Catholic orphanage in Philadelphia, for a series of diagnostic tests in which a tuberculin formula was placed in the test subjects' eyes.[23]

Frustrated in his efforts to control the expansion of vivisection, Dr. Albert Leffingwell, an early twentieth-century physician and social reformer, led a long and energetic campaign to establish restrictions on experimenting with both animals and humans. He pleaded with his medical colleagues to police their actions and establish humane guidelines. "At the beginning of a new century," he wrote in one of his pamphlets, "we are confronted by great problems. One of them is human vivisection in the name of scientific research. We appeal, then, to the medical press of America to break that unfortunate silence which seems to justify, or at least condone it. Now and hence forth, will it not join us in condemning every such vivisector of little children, every such experimenter upon human beings. We make this appeal to it in the name of justice and humanity and for the sake of millions yet unborn."[24]

Unfortunately for reformers like Leffingwell, the scientific current was moving swiftly in the opposite direction. The twentieth century augured a new era of vital, transformative medicine and medical care, and researchers were going to be in the vanguard of it. Innovative treatments, revolutionary preventives, and a whole host of novel procedures were about to explode on the scene, and those preaching restraint, moderation, and prudent, ethical decision making were about to be swept aside as nagging eccentrics and hysterical obstructionists holding back progress. Civilization's advance, along with a healthy dose of journalistic encouragement, fostered an atmosphere that celebrated scientific achievement and medical breakthroughs with no regard for those at whose expense the triumphs were attained.

Each new discovery—a mosquito's role in spreading plague, a certain bacterium doing the same, or a vaccine that could nullify the threat of both—was heroically trumpeted as a great advance. Newspapers, magazines, books, radio, and films repeatedly underscored the significance of the achievements. Medical triumphalism had taken hold of a hero-obsessed culture, and little thought was given to those caught in the wake of the latest scientific success. Those languishing in institutionalized settings—asylums, hospitals, orphanages, and prisons—would make their contribution as well. But they would never see their names in books or newspapers, and they would never earn commendations or receive lucrative endorsements—they were the grist used by the increasingly well-oiled medical mill to achieve the doctors' personal and corporate goals.

There would be no Paul de Kruif to chronicle the ordeal of inhabitants at the New Jersey State Colony for the Feebleminded at Vineland, the Ohio Soldiers' and Sailors' Home Orphans Home, Letchworth Village Colony for the Feebleminded in Thiells, New York, or the dozens of other institutions around the nation that opened their doors and turned their wards over to investigators in white lab coats as so much cheap research material. Doctors had discovered a use for those locked away in state institutions. Children—mute, unwitting, and desperate for affection—would increasingly become the front-line troops in the medical community's quest to improve the world.

# TWO

# EUGENICS AND THE DEVALUING OF INSTITUTIONALIZED CHILDREN

## *"The Elimination of Defectives"*

**DURING THE SPRING AND SUMMER OF 1929, THREE OF THE** nation's most eminent scientists would plan and carry out the deliberate castration of a thirteen-year-old "Mongoloid dwarf" at a state asylum for the "feebleminded" just outside New York City.

The nontherapeutic operation was led by Charles Benedict Davenport, America's leading eugenicist and better-breeding advocate whose indefatigable efforts helped usher "pure prejudice into the stately corridors of respectability."[1] Davenport was assisted in the delicate surgical procedure by the renowned George Washington Corner, a Johns Hopkins Medical School graduate, and by cytologist Theophilus H. Painter, a Yale-trained scientist, who subsequently analyzed the tissue culture.[2]

The use of vulnerable institutionalized children for exploratory procedures, investigative treatments, and experimental preventives was common in the 1920s. America's leading eugenicists like Davenport branded those imprisoned in poorly funded and inadequately staffed county and state asylums, hospitals, and orphanages as devalued members of the tribe, reviled cast-offs from the human race and therefore eminently worthy of study for their peculiarities. Some in the eugenics community found them equally useful as "material" for scientific research—research that was not necessarily related to their conditions.

With the passage of years, institutionalized children would be viewed as increasingly expendable and much sought after as test subjects. World War II and the Cold War that followed further fostered a need for test subjects. Research "volunteers" and institutions holding physically and mentally challenged children became particularly attractive for their convenience, isolation, and affordability. Many researchers viewed such facilities as a gift, a gift that kept on giving.

Utilitarianism, paternalism, and elitism had long been prevalent in the medical research community, but America's infatuation with the better-breeding movement and the nation's involvement in two international conflicts would add a potent rationale for exploitative research practices. Armed with what they considered solid justification for their actions, some doctors who had sworn to do no harm repeatedly breached that oath. Few, if any, spoke up against the practice.

**FOR MANY YEARS, CHARLES B. DAVENPORT** had been fascinated by all matters of hereditary dysfunction, especially those of the "Mongoloid dwarf." As he told the superintendent of the Letchworth Village Colony for the Feebleminded, the syndrome was of unknown origin and had repeatedly proved impossible to remedy. Similar problems in his studies of "dwarf plants as well as insects" had led Davenport to a possible answer; he increasingly believed "Mongoloid dwarfs might be due to abnormalities in the chromosomal complex."[3]

Now in need of humans to test his hypothesis, he knew exactly where to go for test subjects: an institution that held an impaired population he had spent many years studying. In his letter of request, Davenport informed the colony's superintendent that it was time "to examine the chromosomes connected with the dividing cells of a Mongoloid. The only suitable organ where one is sure to find cell divisions is in the testis. The only way to get the tissue in suitable condition for fixation of the chromosomes so they can be observed is by castration."[4]

Davenport believed that pursuing this scientific conundrum was clearly in order and that a Mongoloid child's castration was warranted "for the sake of throwing light upon this problem."[5] The request was quickly granted; no one in authority, including the child's lone parent, the facility's superintendent, or the researchers, seemed opposed to castrating a child for the sake of increased knowledge.[6]

Toward the end of the nineteenth century and while teaching at the Brooklyn Institute of Arts and Sciences, Davenport established a biological laboratory to study primitive marine and mammal life in a lush, marshy section of Long Island called Cold Spring Harbor. It was while overturning submerged rocks, digging up oysters, and inspecting other forms of flora and fauna that he began to seriously wade into the scientific works and theories of Francis Galton, the creator of eugenics. Davenport's strong attraction to Galton's ideas would have great scientific and philosophical portent on both sides of the Atlantic.

Great things were expected of Galton, who came from the illustrious Darwin clan, but a lackluster academic career put an end to such grandiose expectations. A wanderlust and endless curiosity, however, eventually propelled him to become a world traveler, self-taught geographer, and winner of a gold medal for myriad contributions from the Royal Geographical Society.

It would be in the field of heredity, however, that his innate curiosity about human differences and his love of mathematics would produce his most controversial scientific discovery. Galton speculated that cumulative differences in human variation tended to result in mathematical forms and trends that would allow keen observers to predict, over time and in a given population, the variation of a specific variable, such as height, weight, or intelligence. Moreover, while there would be wide differences among large groups of people on many dimensions, differences among generations of people would be slight. For Galton, heredity transmitted not only physical features, such as eye color, height, and weight, but also mental and emotional qualities. And whether an individual was creative or a laggard was not pure chance; it was inherited. The more Galton thought about inheritance, particularly the inheritance of genius in distinguished British families in the 1860s, the more he came to believe that science and mathematics could discover the keys to breeding and that, once understood and properly managed, the human condition could be forever changed for the better. As he stated simply, "Could not the undesirables be got rid of and the desirables multiplied?"[7]

Like many others during the last quarter of the 1800s, Galton pondered a host of hereditary questions concerning race purification and human perfection. By 1883, the year of the publication of his book *Inquiries into Human Faculty and Its Development,* and after years of considering various names for his new science of heredity, Galton finally settled on a combination of two Greek words, *eu* (good or well) and *genes* (born), to create the word "eugenics." The term would soon take on a life of its own and eventually manifest itself in ways, policies, and practices that its originator would hardly recognize.

According to Galton, germplasm, protoplasm or whatever one called that key element responsible for the transmission of physical and mental characteristics from one generation to another was the central ingredient of heredity. He and others at the time didn't fully understand how hereditary traits were passed on, but they knew that environment and quality-of-life factors had little to no impact if one's germplasm declared one unfit.

Galton claimed that he was not unsympathetic to the plight of such families but insisted that if society was to progress, such elements had to be constrained and eliminated. He did not propose to neglect the sick and the unfortunate, but he did advocate their isolation in an effort to stop the production of families likely to include "degenerates."

The demonizing and devaluing of certain segments of the population during the last two decades of the nineteenth century would grow in both support and legitimacy. Galton emphasized getting the most out of society's best representatives, and he favored placing restrictions on society's least admirable members, such as restricting the marriage license process to prevent genetically flawed unions. Eugenics was too important an issue to be a voluntary affair; government enforcement was necessary. During the early years of the twentieth century, the movement not only took root, it also developed a darker, more coercive aura. Galton's "positive eugenics" stressing biologically beneficial marriages to enhance society's benefactors was quickly being supplanted by a "negative eugenics" stressing dramatically accelerated programs and policies to control, if not eradicate, the genetically unfit—those deemed to be medically infirm, racially tainted, and non-Nordic in origin. Those with inferior genetics were dealt with harshly.

Just a decade after Galton invented the term "eugenics" in 1883, Dr. F. Hoyt Pilcher, a US physician and asylum superintendent, castrated nearly six dozen boys at the Institute for Imbeciles and Weak Minded Children in

Winfield, Kansas. Pilcher's predecessor at the institution had been treating young inmates "who were confirmed masturbators" for years without benefit of a proven remedy. Upon assuming the position of superintendent and after "taking a rational view of the subject," Pilcher began his own search for a "curative effect" or solution that would put an end to the annoying and persistent problem. He believed that he had discovered the answer through surgical intervention, specifically, castration.

Many Kansans were horrified at Pilcher's actions, but the medical community was generally supportive, sometimes aggressively so. Calling the "bitter" attacks on Pilcher partisan bombast and "political campaign thunder," one state medical journal said "there is growing sentiment in the profession in favor of castration" as a medicinal vehicle to stem both disease and crime. The journal article also made the case that there was an immediate need for the public to be "educated up" to this realization.

Other doctors were soon asserting that castration could produce therapeutic wonders for a host of medical ailments and social problems. Dr. Everett Flood informed his colleagues at an American Medical Association convention in Philadelphia in 1897 of twenty-six children confined in an institution in Baldwinsville, Massachusetts, who had been recently castrated. According to Flood, "the first castrations were performed with the idea of preventing masturbation in certain cases where the habit was most constant." One boy, the doctor wrote, "had no sense of shame, besides being a confirmed epileptic and of course somewhat feebleminded." Flood said that parental permissions were attained for the surgeries, which were performed by a staff surgeon.

The outcome of the experimental surgeries was more than satisfactory, according to Flood, since "it has been found in every case that the masturbation has ceased." It was also noted that those castrated had become "fairer in complexion," taken on "additional flesh" and were "more manageable," better groomed, and more truthful. Last, but no less noteworthy, Flood pointed out that "the possibility of their transmitting defects to offspring is of course removed."

The last decade of the nineteenth century produced a number of scientific champions of eugenic castration and sterilization. One of them was Harry C. Sharp, a surgeon at the Indiana Reformatory, who became an outspoken leader of the sterilization movement. Sharp often wrote and spoke about the ever-increasing problem of defective germplasm coursing through American society. Sharp's "heroic method" of ridding the world of defective germplasm was sterilization. He would soon report that he had sterilized forty-two young men, some as young as seventeen. He would go on to sterilize hundreds more in order to ensure that they did not become criminals and an economic drain on society. Delighted with his results, Sharp argued that "radical methods are necessary" and encouraged fellow physicians to press elected officials to have superintendents of insane asylums, prisons, and institutions for the feeble-minded sterilize every male inmate who walked through the front door.

The reaction to Sharp's paper by his peers was generally favorable and supported by asylum superintendents from Cuban leper colonies to Ohio penal facilities. In fact, Sharp would go on to be elected president of the Physicians Association of the National Prison Association, an office from which he could propagandize with even greater authority. Sharp was only one of an increasing number in the medical profession advocating for sterilization as an acceptable therapeutic practice. Articles written by physicians promoting "sterilization for social control" greatly increased between 1880 and 1900. Most followed the typical pattern of decrying the growth in crime and the rising cost of incarceration and proudly extolling patients' new sense of industriousness, ambition, and happiness after surgery.

More physicians were signing on, some of whom were nationally prominent. Dr. William T. Belfield, a surgeon of some stature, endorsed both eugenics and mass sterilization policies in December 1907 at a Physician and Law Clubs of Chicago meeting. His presentation, which was soon after printed in the *Journal of the New Mexico Medical Society,* argued that the American homicide rate was spiraling out of control and already "33 times higher than London."[8] Belfield urged Illinois lawmakers to pass a sterilization law similar to the recent Hoosier state law; others would initiate a similar lobbying campaign

Dr. David Weeks, the head physician at the New Jersey State Village for Epileptics; Dr. Charles Carrington, chief surgeon at the Virginia State Penitentiary; and Dr. F. W. Hatch, secretary of the State Lunacy Commission in California, were all actively writing, lobbying, or utilizing eugenic nostrums to deal with the nation's "social disease." Some eugenic proponents, such as Lewellys F. Barker and G. Frank Lydston, were not just directors of institutions but major names in the field of medicine. Barker was the physician-in-chief at Johns Hopkins Hospital; Lydston, a professor of genitourinary surgery at the University of Illinois, had some extreme views on the role of medicine in combating America's dysgenic hordes—Southern European immigrants—and was not bashful about sharing them. Along with advocating castration of rapists and sterilization of the unfit (epileptics, inebriates, consumptives, etc.), he also proposed the use of toxic gas in sealed compartments "to kill convicted murderers and driveling imbeciles." Barker's and Lydston's papers, especially those published in the *Journal of the American Medical Association* and the *New York Medical Journal,* signified for many Americans that eugenical surgery had been accorded a seal of approval by the profession's elite.[9]

In 1910, nearly two dozen medical articles advocated eugenical sterilization, and between 1899 and 1912, thirty-eight articles were published calling for the profession to be more proactive on the issue.[10] As one proponent of medical intervention would forcefully state, "We must face the fact the very life-blood of the nation is being poisoned by the rapid production of mental and moral defectives, and the only thing that will dam the flood of degeneracy and insure the survival of the fittest, is abrogation of all power to procreate."[11] Clearly, the medical profession regarded criminality, feeblemindedness, and the burden and expense of degeneracy—"the very nightmare of the human race," as one eugenics activist called it—as a social and economic pox that had to be eradicated. Medical support enabled twelve states during that period to enact sterilization laws (four additional states—Pennsylvania, Oregon, Vermont, and Nebraska—passed similar legislation only to have it vetoed by the governors).

With this as a backdrop and nearly two decades later, during the dawn of the Great Depression, Davenport's desire to castrate a child for research purposes was met with something approaching casual indifference by the asylum superintendent.

Interestingly, the nascent eugenics movement would find its most ardent supporters in America, and they would champion a panoply of harsh and restrictive policy initiatives including segregation, deportation, castration, marriage prohibition, compulsory sterilization, and occasionally euthanasia or withholding medical treatment to those in need. Among such proponents, the issue of medical experimentation on "defectives" and other unfit "material" would not be viewed as a departure from normal policy and standard medical procedure.

Though many prominent Americans such as Alexander Graham Bell, Theodore Roosevelt, Luther Burbank, and Margaret Sanger would at one time or another gravitate to the eugenics cause, one of the earliest, most devoted, and hardest-working of Galton's disciples would prove to be Charles Davenport.

Negative eugenics and the prevention of the proliferation of tainted genetic material became Davenport's sole concern. The more quickly and aggressively America addressed the challenge, the better it would be for the nation's collective health and bloodlines. In Davenport's mind, the country was being overrun by an array of defective elements. Viewed above all as a potential and actual threat were the "feebleminded"—a catchall term of the era that included a wide swath of mental symptoms and syndromes—who made up the bulk of any community's motley collection of mentally challenged beggars, thieves, lunatics, and worse. Those with severe cases of mental or physical retardation were obviously in this category, but also swept up in the negative eugenic dragnet were those who spoke poor English, who stuttered badly, or who were just painfully shy around other people, regardless of their actual intelligence and abilities. Davenport believed such undesirables made up 10 percent of the general population and often referred to them as the "submerged tenth."

As one ardent eugenicist complained of the unfit in 1935, "the vast sums in taking care of lower grades of our feeble-minded people . . . will necessitate our spending more and more."[12] Something needed to be done that would "work in the elimination of undesirable elements in society."[13] Some, like Alexander Johnson, a former secretary of the Indiana Board of State Charities in the 1930s, advocated "segregation." Johnson believed "all such people should be taken in childhood . . . and placed in a training school . . . where they may be placed in useful labor, and kept under positive control as long as they live."[14] Many others, however, believed that the nation's orphanages, prisons, and mental asylums were already overcrowded and that a more radical approach was needed to rectify the problem.

For eugenics activists like Leon Whitney, "elimination" was the answer. "If in one fell swoop," he argued, "we could eliminate all our useless degenerates, incapable of anything beyond a kind of gross animal happiness, if we could awaken one morning and find all these gone in some mysterious but painless fashion, what class of persons would we fix on to be the ones eliminated?" Whitney already knew the answer. They should "start eliminating at the bottom. We should go to the institutions for the feeble-minded and look at their inmates. The first ones to be picked out would probably be those of so low grade as to be hardly better than human vegetables."[15] Stamped with an array of mental and physical stigmata more visible and damning than any scarlet letter, "defectives" languishing in large state institutions were increasingly viewed in the early decades of the twentieth century as a public menace and a financial encumbrance requiring swift action.

Many prominent citizens joined the condemnatory bandwagon. For example, in 1913, former President Theodore Roosevelt wrote Davenport, "I agree with you . . . that society has no business to permit degenerates to reproduce their kind. . . . Some day, we will realize that the prime duty, the inescapable duty, of the good citizen of the right type, is to leave his or her blood behind him in the world; and that we have no business to permit the perpetuation of citizens of the wrong type."[16] Margaret Sanger, the great social activist and birth control proponent, was even more strident in her denunciation of society's unfit elements, "vigorously oppos[ing] charitable efforts to uplift the downtrodden" and arguing that "it was better that the cold and hungry be left without help" so the eugenically fit would face less of a challenge from "the unfit." She often compared the poor and the great mass of dispossessed to "human waste" and "weeds" needing to be "exterminated."[17] Other presumably progressive citizens like G. Stanley Hall, Washington Gladden, and E. E. Southard—all established academics and respected citizens at the time—were similarly disposed.

Davenport believed that many of the immigrants from Europe and Asia who were entering the country were either damaged or dangerous, occasionally both, and were destined to prove an additional financial burden. Davenport was already furious that the government—and therefore the taxpaying public—was obligated to support the growing number of people in almshouses, asylums, and prisons. Many tens of millions of dollars were involved in their maintenance, and there seemed to be no end in sight.[18] With the European and Asian floodgates open, America had to protect itself. Careful observation and study could detect the most menacing of the unfit arriving on our shores. Even a cursory examination of an individual or a group's lineage could determine their social inadequacies or criminal tendencies, for, Davenport believed, there was such a thing as "racial identity," and it determined behavior. Poles, for example, were "clannish"; Italians tended toward "personal violence"; and Jews, though engaged in little "personal violence," were prone to "thieving" and "prostitution."[19]

Davenport was as much a social activist as a scholarly academic, and he targeted nearly a dozen "socially unfit" groups for "elimination." They included the poor; alcoholics; all classes of criminals; epileptics; the insane; the weak; those subject to specific diseases; the physically deformed; and the deaf, dumb, and blind.[20] Though frowned on and shunned by polite society, many of these classes were soon to be sought out as ideal research material.

No novice as an organizer, Davenport realized that such a battle plan required economic resources. His organizational and entrepreneurial skills would serve him well over the years, but never more so than in his 1909 meeting with Mary Harriman, the daughter of one of the nation's most prominent railroad magnates and someone who was already familiar with the Cold Spring Harbor Laboratory from her days as a Barnard College undergraduate.

A liberal and social activist, Mary Harriman was drawn to the "social improvement" aspect of Davenport's pitch, as were many progressives of the day, and she encouraged her mother to support the young Brooklyn scientist's ambition to establish a serious eugenics research center. Armed with facts, a clear vision of what needed to be done, and the commitment of a true zealot, Davenport proved an impressive and convincing advocate. He received an endowment from Mrs. E. H. Harriman that enabled him to purchase a seventy-five-acre tract of land just up the road from his Cold Spring Harbor Laboratory. The Eugenics Record Office (ERO), with sufficient funds and staff to carry out its mission of racial hygiene and endless file cabinets to hold an ocean of extremely sensitive personal information, would become the largest repository of hereditary data in the nation. It would also be home to an army of field investigators who would make thousands of house calls and official visits to asylums, prisons, and hospitals across the country and ferret out information about inmates and their "defective lineage." Davenport saw the Harriman gift as a turning point in his eugenics campaign and referred to it in his diary as "A red letter day for humanity."[21]

By 1918 Mrs. Harriman would give over half a million dollars to various eugenic causes, a considerable sum in those days. Davenport made sure that a portion of the money went to his front-line troops: young women from elite colleges such as Vassar, Radcliffe, and Wellesley and young men from universities such as Cornell, Harvard, and Johns Hopkins who had a knowledge of biology and went into the field to spread the word and seek out information about the unfit in America.

With an information-gathering army approaching 300 trained field workers, Davenport not only had the means of spreading his alarmist vision of national decay and impending doom, but he was also able to collect data, family profiles, and statistics on everything from instances of pellagra and nomadism to an individual's athletic proclivities and emotional disposition. Most, if not all, of the information was used in an array of books, bulletins, pamphlets, and newsletters like *Eugenical News* supporting sterilization, a strict immigration policy, and other efforts to enhance American germplasm and thwart degeneracy.

Shortly after procuring the Harriman money, Davenport obtained an even larger Rockefeller Foundation contribution. Davenport and the ERO were now forces to be reckoned with. Money made all the difference for the eugenics movement. Biological supremacy, raceology, and coercive public policies were just excess verbiage until those notions were backed by hard cash.[22] Money now greased the skids for simplistic hereditary theories to become law. The monetary windfall enabled Davenport to turn his data-collection system from a relatively tiny cottage industry into a statistical powerhouse that couldn't be easily ignored. And by recruiting asylum superintendents to the cause, traveling to their institutions, measuring and examining the shuttered "defectives," and theorizing about the origins of degeneracy and the best methods to combat its insidious growth, he established a formidable template on how to tackle the nation's problem.

Moreover, America's growing number of asylums proved a gold mine of investigative opportunity that held "the ore of progress," according to E. E. Southard, a Harvard Medical School neuropathologist.[23] Scientists were discovering the great opportunities presented by large state institutions as human laboratories. The unfit had finally assumed some value as laboratory specimens. Davenport and other eugenics adherents immediately recognized the possibilities and became familiar figures on institutional campuses. Letchworth Village, a new facility for the "feebleminded" where Mary Harriman served as a member of the board, became one of his favorite venues. Designed to "provide training and education for backward children," Letchworth Village, located just north of New York City, trumpeted its goal of "increasing better human strains . . . and the right to be well-born."[24]

Davenport's legwork paid dividends as many institutional heads bought into the campaign and bid for his favor. He thanked Henry H. Goddard, the research director of the New Jersey Training School for Feeble-Minded Boys and Girls, for his efforts to hire field workers to research "the pedigree of feeble minded children." He was "quite convinced" that the best method for procuring the necessary information was "to visit the families of your inmates in their own homes."[25] It would not be long before Davenport's ERO field workers were armed with a five-page questionnaire that asked dozens of personal questions, ranging from one's religious habits, marital history, and use of tobacco and alcohol, to a list of illnesses since infancy, surgical procedures, and those in the family who were cross-eyed, had "fingers with only 2 joints," and had "abnormal genitalia." Intrusive inquiries were also made regarding those in the family who were guilty of "deceitfulness," "cunning," "revengefulness," and "masturbation."[26]

Some viewed the accumulation of such salacious personal information as potentially explosive, especially if it were to "fall into the hands of newspaper or magazine writers who would be very glad of an opportunity to exploit the matter," as Goddard wrote to Davenport. It was one thing for readers of *American Breeders Magazine*—which was devoted to doing to humans what cattle farmers did to their livestock—to glean information derived from the research of ERO field workers; it was a whole other thing if the mainstream press gained control of it. Goddard's "cautionary" note to Davenport suggested that the data were "in danger of being used by some one who will put it into a popular magazine and thus queer the whole country, so to speak, in regard to this whole matter of Eugenics."[27]

Undaunted, Davenport admitted that it was possible that "yellow journals might succeed in using this material to put such an aspect upon all our whole work as to damage it seriously. There are scores of persons" in the newspaper business, he conceded, "who would not scruple to use their power to damage any good cause." And finally, he argued boldly, "There is no use in collecting facts unless they can be put before at least the scientific public to form the basis of eventual action."[28]

Goddard would make his own contributions to eugenics history. One of them was the influential tract *The Kallikak Family: A Study in the Heredity of Feeblemindedness*, which traced the origin of an eight-year-old female Vineland resident back many generations to a Revolutionary War soldier who fathered a defective son after impregnating a feebleminded girl. Despite its questionable conclusions and lack of sophistication, Goddard's research on "mental defectives" would become increasingly important to Davenport and the eugenics cause.[29] As the relationship between the men grew, they would regularly notify each other of new developments and discoveries. Always on the lookout for interesting examples of unfit germplasm, Goddard and Davenport shared stories of degenerates and inbreeding they had come upon in their travels.[30]

Though Davenport may have had an academic interest in isolated instances of degeneracy, he was more concerned about winning converts, building a movement, and spreading the gospel of eugenics. Getting the message into high schools and colleges in the 1910s and 1920s was a major preoccupation, as was keeping track of what biological material was being taught, what eugenics courses were being offered, the number of students in attendance, and the total class hours devoted to "laboratory" or "fieldwork." Davenport had the ERO and the Carnegie Institution of Washington blanket the nation's universities with questionnaires, and a surprising number of institutions replied. At Adelphi College in Brooklyn, for example, 51 females were enrolled in a seventeen-week eugenics and human heredity course; at Brown University in Providence, 75 students—45 males and 30 females—participated in a four-week course for upperclassmen; 88 males at Dartmouth College took a sixteen-week course in animal and plant heredity that included a strong eugenics component; and 187 students at Penn State were learning eugenic theory in a seventeen-week Botany and Genetics course.[31]

Some universities, like the University of California, were even offering "extension courses" to the public that allowed both students and average citizens to be schooled in the new science during a "fifteen lecture" program at the San Francisco Main Library. Lectures would present a smorgasbord of eugenic thought, including "The Transmission of Human Defects," "War and Race," "The Elimination of Defectives," and "Sterilization and Segregation."[32] Professional degree programs were very much part of the propaganda mill associated with eugenics.

The Pediatrics Department of Hahnemann Medical College in Chicago, for example, reported that it had sixteen third- and fourth-year students enrolled in a thirty-four-week eugenics course. Hahnemann was not unique; numerous medical schools were propagating eugenic nostrums for the physical and mental maladies plaguing the body politic. Adolf Meyer, an early and active eugenics supporter, appreciated Davenport's taking time out from his busy schedule to visit with Johns Hopkins medical students. "I think it is an excellent thing to make it possible for those starting on medical work to have a correct and real conception of what your Institute is doing," Meyer wrote Davenport. "I am making a very determined effort to rouse a working interest . . . with quite a little response."[33]

It should be pointed out that not every school of higher education subscribed to this class and race warfare, not to mention the outright bigotry that many eugenicists regularly exhibited. As one college official candidly informed the ERO superintendent, "Allow me to say very frankly that in the opinion of the faculty of the University of Detroit, Eugenics is full of dangers. . . . My desire is to protest against the brutal and superficial heredity terrorism with which certain modern eugenic enthusiasts advocate a regulation of human breeding borrowed from the stables and totally foreign to the human race. . . . Moreover, it is only in the rarest cases that we find two parents who are both of them, physically and psychically, so equally and heavily tainted and defective, that anything could be safely predicted with regard to their children."[34]

There were other critics, some with national reputations, such as former presidential candidate William Jennings Bryan and influential newspaper columnist Walter Lippmann, but their pointed criticisms were occasionally lost in the enthusiastic chorus calling for implementation of this new science promising a quick journey to a progressive utopia. Lippmann, for example, had numerous qualms about many eugenic tenets, including the movement's utopian vision. He often voiced his distaste for eugenics, especially its increasing devotion to IQ testing during and after the First World War. Expressing his dissatisfaction in the pages of the *New Republic* in 1922, he soberly argued, "The whole drift of the propaganda based on intelligence testing is to treat people with low intelligence quotients as congenitally and hopelessly inferior." He did not accept the principle that a person's capacity to learn, mature, and develop was "fatally fixed by the child's heredity."[35] Lippmann would subsequently admit his passionate disapproval of much in the eugenics hymnbook by writing, "I hate the impudence of a claim that in fifty minutes you can judge and classify a human being's predestined fitness in life. . . . I hate the abuse of scientific method which it involves. I hate the sense of superiority which it creates, and the sense of inferiority which it imposes."[36]

Fortunately for the eugenicists, the disapproving tone of Michigan Jesuits and political pundits was not often replicated at the collegiate level; most institutions of higher learning either fell in line and offered eugenics-based or related courses or remained silent. Many of the nation's premier academic institutions offered eugenic theory and curriculum in one form or another. Prestigious universities—Harvard, Columbia, Cornell, and the University of Chicago—had top faculty teaching such courses. Large state universities like Wisconsin were also on board; Northwestern was said to be "a hotbed of radical eugenic thought." Princeton had Harry Laughlin, a key ERO operative, teaching a course, and E. G. Conklin, Princeton's president, had beseeched Davenport as early as 1909 to "lecture before [the school's] Natural Science Club."[37] V. L. Kellogg, a noted entomologist, taught a course on eugenics and zoology at Stanford, and top academics like Herbert Jennings and Raymond Pearl offered eugenics courses at Johns Hopkins.[38]

In fact, "eugenics rocketed through academia, becoming an institution virtually overnight," according to one observer of the movement. By the start of World War I, over forty-four major universities were offering eugenics courses. Less than a decade later, that number would mushroom "to hundreds, reaching some 20,000 students annually."[39] Eugenics would become so commonplace on American college campuses by the mid-1930s that Leon Whitney, secretary of the American Eugenics Society, proudly claimed, "Eugenics is being taught now in three-quarters of our five hundred colleges and universities, and in many high schools and preparatory schools."[40]

The upshot is that at least two generations of college students—the future lawyers, teachers, doctors, and medical researchers of America who had sworn to do no harm—were being imbued with an air of superiority and a condescending disdain for the weak, the infirm, and the institutionalized of the nation. If eradication of such appalling and burdensome elements was a national priority, then their periodic use as research material for society's benefit was a worthy, even commendable act.

Some colleges became so steeped in this new quasi-religious faith in science that the campuses seemed saturated in an all-encompassing eugenic haze. The "unfit" in combination with occasional racial turbulence were often seen as the reasons for political unrest, social turmoil, and public disenchantment. America, particularly in the South, was no longer a harmonious "melting pot" but a boiling cauldron of social discontent. Some professors at elite southern schools were even predicting "racial suicide" if society's best families didn't reproduce more progeny and if legislative action wasn't taken more quickly on the social and legal front. The University of Virginia, then the largest institution of higher education in the South and an academic model for the region, was one of these bastions of hereditarianism and eugenic thought.

The Commonwealth of Virginia would champion the legality of sterilization during the post–World War I quandary concerning the constitutionality of the nation's many sterilization statues. Eugenic activists hoping to take the issue to the Supreme Court believed they had found a suitable vehicle in the form of a seventeen-year-old girl named Carrie Buck. Classified a "moral imbecile," Buck, then an inmate in the Virginia Colony for Epileptics and Feebleminded in Lynchburg, had given birth to a child out of wedlock just prior to her placement in the institution.

Emma Buck, Carrie's mother, was also a resident there. She had been certified a "moron" with the mental age of an eight-year-old and had developed the reputation as a prostitute and one of Charlottesville's least respected women. Rumors circulated that she was a "feebleminded" alcoholic and an unfit mother. Virginia authorities believed that if Vivian, Carrie's infant daughter, showed signs of feeblemindedness, that would document three successive generations of mental deficiency, a perfect test case for the Supreme Court to rule on. To underscore their argument, Virginia authorities consulted with Harry Laughlin, Davenport's chief deputy at the ERO. A zealous, indefatigable eugenicist, Laughlin was instrumental in the recent passage of federal legislation greatly limiting the number of immigrants entering the United States. He would be instrumental in fostering similar eugenics public policy remedies in the courts.

Though Laughlin brainstormed with Virginia authorities about legal nuances, helped plot strategy, and examined numerous documents, he never actually met any of the women involved in the case. Despite that glaring omission and conflicting information about the actual mental impairments of those at the center of the controversy, he enthusiastically confirmed that the Buck women "belong to the shiftless, ignorant, and worthless class of anti-social whites of the South." Other aspects of the case raised similar concerns from a scientific and judgmental point of view. Young Vivian, for example, barely seven months old at the time, was said in the testimony of one critical Red Cross worker to have a "look" about her that was "not quite normal."[41]

Despite prophetic warnings from defense counsel and a court consisting of such renowned members as Louis Brandeis and former President William

Howard Taft, the vote was 8 to 1 in favor of the Virginia statute. In the court's now-infamous opinion in *Buck v. Bell,* Justice Oliver Wendell Holmes wrote, "We have seen more than once that the public welfare may call upon the best citizens for their lives. It would be strange if it could not call upon those who already sap the strength of the State for these lesser sacrifices . . . in order to prevent our being swamped with incompetence. . . . The principle that sustains compulsory vaccination is broad enough to cover cutting the Fallopian tubes." Holmes followed with the infamous line, "Three generations of imbeciles are enough."[42]

For those devalued souls already interned in state asylums, the decision would usher in a eugenic tsunami. An ever-increasing number of children would now be called on by both the state and medical investigators for these and other "lesser sacrifices."

After several decades of persistent and pervasive eugenic propaganda and at least twenty years of intense personal organizing around the subject, Davenport had achieved considerable success, and his request to castrate a "Mongoloid dwarf" in 1929 provoked little alarm.

Davenport and his Cold Spring Harbor research facility were known to asylum superintendents around the country. C. S. Little, the superintendent at Letchworth Village in Thiells, New York, was certainly familiar with him and his group's work. The two men had been corresponding for years, and Davenport and his "trained" field workers were so often on asylum grounds "measuring" and "examining" and "studying" various defective populations that many must have thought them regular staff members.

The opportunity Letchworth Village provided to his research team was not lost on Davenport. Like scores of researchers who would flock to such institutions in coming decades for access to test "material," he often expressed his gratitude. Davenport knew how to flatter the egos of those who were the gatekeepers of the charges he wanted to study. As he told Little in one missive, "I was glad to have [field workers] get acquainted with you and be introduced by you into the Village into which you have woven so much of your personality."[43]

Davenport's interest in feebleminded children was all-consuming. All matters related to their mental and physical nature interested him. In a 1927 letter to Little, he notes that he has collected the "physical measurements of 100 boys during three or four years" in an effort to study the "relations between physical and mental development," but he now wanted to compare the results to those of "mongoloids in other institutions." Davenport closed by saying he would like "to spend an additional day each month for as many months as may be necessary to carry out this program."[44] As happened repeatedly over the many years of their relationship, Little wrote back to the nationally recognized eugenicist that "we shall be glad to have you come anytime if you can find material to work with."[45]

Over time, Davenport would go on to help create and lead the Letchworth Village Research Council, which would "investigate the causes . . . that prevent the full development of the individual, making him a financial and social burden to the community." The result of such a council of experts, argued Davenport, would "bear fruit eventually in lowered public expense and greater public security."[46]

Davenport told one Letchworth Village official that correcting the problem of the feebleminded depended on efforts "to dry up the springs from which they arise. The fact that from defective parents defective offspring arise" made it necessary to implement numerous precautionary measures, including mass sterilization programs. One in particular was "to sterilize the girls before they were discharged from the institution." In fact, argued Davenport, the "solution seems to be the passage of a general sterilization law and its effective demonstration."[47]

Doctors and asylum bosses held equally expansive views on experimentation; many believed scientists deserved a free hand, especially in the use of the institutionalized. As medical historian Paul Lombardo states in his article on the castration of a "mongoloid dwarf," Superintendent Little "did not think it necessary to get written consent from one candidate's mother." Davenport, however, was concerned about the "law" and potential "legal" issues and preferred some form of parental consent. Deciding to take the path of least resistance, they passed on one "mongoloid" child whose parent was too "intelligent" and selected another whose mother was "less capable" intellectually to discern what was being orchestrated. Despite her agreement, it was feared "her permission would probably have no legal standing." To ensure that a "case could be made" for performing the surgery, Davenport pushed the rationale that the procedure was being done on "therapeutic grounds." The boy, they would argue, exhibited "a rather marked eroticism and this probably bothers him some, as it doubtless does his attendants."[48] Like many other such institutions, Letchworth Village repeatedly proved itself a valuable incubator for myriad medical studies.

There can be little doubt that the American eugenics movement exacerbated an already grave, if not intolerable, situation for those with acute mental and physical challenges. Discarded, shunned, and often the brunt of pranks, jokes, and more serious criminal acts, the "feebleminded" and physically challenged were destined for compromised, difficult lives. The eugenics movement of the early twentieth century would show them no mercy. Proponents of "better breeding" wanted more than separation and institutionalization for such dysgenic individuals; they wanted to once and for all eliminate the unfit from the nation's gene pool: in short, to finally be rid of them.

Doctors and medical researchers dedicated to seeking answers for the numerous health conundrums confronting mankind—especially the more zealous among them—instinctively knew what to do with such a loathed and flawed population. Its members could serve humanity and science by participating in important research studies, experiments designed to increase knowledge and help solve long-standing medical mysteries. With the passage of years, children, and "defective" children in particular, would prove increasingly attractive as test subjects, and those in a central repository like an asylum or orphanage were an experimental gold mine.

The gradual but steady trend of utilizing institutionalized children as cheap and available research material would prove surprisingly easy. Recruitment of minors as test subjects for experimental purposes had long been a part of medicine, but the practice grew more frequent with the onset of the eugenicists' sterilization and castration campaigns. Asylum children were chosen as test subjects precisely because they were isolated, out of public view, and considered of less value than normal children. In these isolated venues, experiments that resulted in death or some other unfortunate outcome could more easily be kept from the general public.

In 1895, between Drs. Sharp and Belfield's clarion calls for greater utilization of surgical sterilization, Dr. Henry Heiman notified colleagues at the New York Academy of Medicine that he had purposefully "inoculated" a four-year-old boy "suffering from idiocy and chronic epilepsy" with "pure cultures of gonococci."[49] In a similar experiment, a second boy "suffering from idiocy" was subjected to an inoculation of gonococci; like the other boy, he experienced the nagging and painful symptoms of gonorrhea.

Those attending the New York medical forum were said to be "much interested" in Heiman's research and went out of their way to complement "his painstaking work." Discussants contributed numerous thoughts and comments; none concerned the ethics of using intellectually challenged, institutionalized children as test subjects. In fact, such practices would become more common at the turn of the century, and those choosing to use institutionalized populations were viewed by their peers as innovative, judicious in their use of available resources, and dedicated to their craft.

Many physicians and medical researchers not only bought into this line of thinking but became vigorous proponents of it. With their elevated professional status, men of science would come to believe that their decisions on such matters as human experimentation were not only sound and prudent but also ethically unassailable. They were the gods of science, omniscient arbiters of public health, oblivious to their fierce desire to pursue new medical theories and methodologies at the expense of ethical boundaries related to patient inviolability and the line between therapeutic and nontherapeutic experimentation. Regardless of whether it was the doctor at the bedside or the researcher in the lab, training combined with deeply embedded paternalistic and elitist beliefs told them they knew best. Medical journal editors celebrated their discoveries, popular media praised their heroic contributions, and the public applauded their triumphs. The few critics bold enough to protest such practices, such as the antivivisectionists, were easily marginalized. The longer those practices went unchallenged by significant members of the community, the more accepted and widespread they became.

That pattern would continue until the 1940s, when the use of the institutionalized as human guinea pigs would shift from a relatively small cottage industry to a full-blown, government-funded, university-sponsored pillar of modern medicine.

# THREE

# WORLD WAR II, PATRIOTISM, AND THE NUREMBERG CODE

*"It Was a Good Code for Barbarians"*

**FROM DECEMBER 1946 TO AUGUST 1947, TWENTY-THREE** Nazi doctors and medical administrators were put on trial at the Palace of Justice in Nuremberg, Germany for "murder, tortures and other atrocities in the name of medical science." Officially known as the *United States of America v. Karl Brandt et al.*, the landmark historical event is also known as "the Doctors' Trial." Accomplished physicians and high-ranking Nazi officials such as Karl Gebhardt, Victor Brack, and Wolfram Sievers were put on trial for orchestrating experiments on concentration camp prisoners where victims were immersed in icy water, forced to swallow seawater, enveloped in toxic gas, placed in vacuum chambers, injected with plague, and subjected to horrible bone transplantation surgery. Ultimately, seven Nazi defendants were sentenced to be hanged, and many others received long terms of imprisonment. The Doctors' Trial concluded with the court tribunal handing down a list of ten principles—the Nuremberg Code—designed to formulate a universal standard for the protection of human subjects involved in medical research. Regrettably, that high standard of human rights has been difficult, if not impossible, to achieve.

Most medical professionals in industrialized nations, particularly those in America, considered the Nazi experience a medical aberration, an anomaly not indicative of how the rest of the civilized world practiced scientific research. Columbia University medical historian David J. Rothman, for example, has argued that "madness, not medicine, was implicated at Nuremberg." The "prevailing view" at the time and thereafter, according to Rothman, was that those on trial at Nuremberg were "Nazis first and last; by definition, nothing they did, and no code drawn up in response to them, was relevant to the United States."[1]

Moreover, argues Rothman, Nazi medicine was interpreted by some as an example of the evils of state intrusion. The Third Reich's "atrocities were the result of government interference in the conduct of research. Science was pure," these critics claimed. "It was politics that was corrupting." The upshot for the 1947 Nuremberg Code was clear; rather than being seen as a model for the regulation of human experimentation, it was increasingly becoming viewed as an argument against "socialized medicine" and in favor of the position that "the state should not interfere with medicine."[2]

Interviews with numerous physicians have confirmed that the Nuremberg Code was rarely, if ever, mentioned as part of their medical education during the 1950s and 1960s. It certainly was not the centerpiece of any ethical training regarding the limitations on research with humans. In fact, an argument can be made that those few aware of it were concerned about the code's numerous restrictions on human research. The most knowledgeable, it would appear, were in search of ways to loosen, if not jettison, the many limitations that the code's ten principles imposed on scientific investigators. It was because of this entrenched and uniform opposition, according to Yale professor and ethicist Jay Katz, that the Nuremberg Code "was relegated to history almost as soon as it was born."[3] In less than a decade and a half, those disaffected elements became a concerted movement that resulted in a competing set of principles, ones much more to the liking of physician-investigators.

The Declaration of Helsinki adopted by the World Health Organization in 1964 was a document drafted by doctors, for doctors, and with doctors' interests in mind. "Medical progress" now superseded the "interests of the subject" in the hierarchy of principles. Gone was the Nuremberg Tribunal's strident language prohibiting "force, fraud, deceit, duress, overreaching, or any ulterior form of constraint or coercion," as was an earlier Helsinki draft provision stating that "prisoners of war, military or civilian, should never be used as subjects of experiment." Thought injurious to the unfettered research arena that physician-researchers desired, the more restrictive provision was excised from the document. The new Helsinki Code trumpeted "the advancement of science" over "the integrity of the individual."[4]

Further bolstering this philosophical shift in emphasis were the US government's rules and regulations for the protection of human research subjects, which confirmed the medical profession's efforts to distance itself from Nuremberg. The result, not surprisingly, was a postwar research landscape littered with the damaged bodies and embarrassing ethical transgressions of physicians and institutions that felt comfortable taking liberties with a code of research conduct never truly embraced by the nation that formulated it. Widely known cases of unethical medical research, such as the Tuskegee syphilis study and those occurring at the Jewish Chronic Disease Hospital in Brooklyn, the Willowbrook State School on Staten Island, and Holmesburg Prison in Philadelphia, are just four examples of vulnerable populations being exploited in the aftermath of the Nuremberg Trial, but many others will be identified in coming chapters. If tenets of the Nuremberg Code had been respected, there would be far fewer instances of unethical medical research in America.

The question naturally arises: How could American doctors, presumably the best and brightest of the medical profession—and according to Nuremberg medical expert Andrew Ivy, the most ethical—have committed such egregious acts on vulnerable populations at home and abroad at the very same time that representatives of the profession were lecturing Nazi doctors about how medical research should be conducted? Such hypocrisy is not surprising; American researchers had been incorporating vulnerable subjects in their work for some time. That practice had accelerated as World War II placed an increasing burden on doctors to discover a raft of cures and preventives for a host of diseases and medical conundrums. A slippery slope of research convenience and opportunism had been created where need and immediacy trumped ethics. Reining in such an entrenched institutional culture would take more than a distant trial of "barbarians" and the creation of a novel but unwieldy code of ethics. The last thing researchers wanted was a code of ethical conduct unsuited to a supercharged American medical community that was about to flex its institutional and entrepreneurial muscle further than anyone could have anticipated.

**DURING THE EARLY YEARS OF THE WAR,** American medicine, and human research in particular, took the giant step from a relatively laid-back cottage industry to a well-funded, high-powered machine. As medical historian David Rothman writes, "The transforming event in the conduct of human experimentation in the United States, the moment when it lost its intimate and directly therapeutic character, was World War II."[5] A landscape dominated by unsophisticated, underfunded, and solitary practitioners striving to solve medical riddles quickly evolved into well-funded, highly coordinated scientific campaigns geared toward helping troops on the front lines. Success of the mission dominated the campaign; the war's urgency subordinated all other factors, including the issue of consent. Test subjects were increasingly at risk.

Embroiled in a life-and-death struggle with ruthless totalitarian powers across the globe, the threat of annihilation dictated a unified effort, individual self-sacrifice, and contributions by all, even those on the home front who were sequestered in custodial institutions. Criticism regarding the use of institutionalized citizens as subjects for research was nonexistent. A war had to be fought and won, and all were expected to contribute to the effort.

Those warehoused in state institutions—from penal facilities to mental asylums to orphanages—were sought after by researchers in need of subjects to test new medicinal potions. Letters between doctors that began "I am writing to you with regard to our conversation about the possibility of using patients of certain institutions as subjects for trial of new vaccines against . . ." were commonplace during the war.[6] War Department officials concerned about US servicemen contracting everything from measles, dysentery, and influenza to more exotic biological threats such as sandfly fever, dengue fever, and scrub typhus fever were desperate to discover appropriate treatments and preventives as quickly as possible. But that required experimentation; integral to any successful research effort was the ability to procure a sufficient number of test subjects to produce a disease-fighting elixir.

Research physicians were always on the hunt for test subjects. The acting director of the Commission on Neurotropic Virus Diseases, for example, sent out numerous missives stating, "I am inquiring whether it might be possible to find 1,000 to 1,500 candidates in the various State institutions with which you are familiar." The letter went on to underscore that the "problem . . . is intimately and directly associated with the War effort," and the government would be most "grateful indeed if there is anything you can do to help us in this matter."[7]

The letter was sent to, among others, Dr. Joseph Stokes Jr., a highly regarded physician and pediatrics researcher at the Hospital of the University of Pennsylvania and the Children's Hospital of Philadelphia. Stokes and many top medical scientists like him with access to those warehoused in state institutions generally complied with such requests. Long familiar with the drill, Stokes had been testing new theories and perfecting medical remedies years before the war began. A longtime advocate of the immunity-boosting properties of gamma globulin, Stokes was absorbed by the latest developments in fighting childhood diseases. Everything from influenza to polio was of interest, and he did more than his fair share of investigational research. But with each new frontal assault on a particular malady came the need for test subjects. Stokes knew where to go for cheap and available "material."

As Stokes informed medical writer Paul de Kruif, "permission has practically been granted for an extensive experimental study . . . in an institution of 700 inmates, ranging (in age) from five years upwards." Such facilities, Stokes assured de Kruif, were not in short supply, "and if we fail to get permission from this institution, there are other groups which we can use."[8]

Institutionalized children weren't necessarily researchers' first choice as experimental subjects, but complications anywhere along the research paradigm could short-circuit a project or speed up an investigation before its time and quickly find its way into a school, orphanage, or facility for "feebleminded" youngsters. For instance, in a subsequent series of correspondence, Stokes would inform de Kruif that he was "getting much disturbed about the situation . . . with the monkeys," whose cost was rising precipitously. "The price of monkeys," he complained to his colleague, was once again "going to be increased." If "additional funds" were not procured from some source, he warned, work would stop. Or in some cases, researchers would either resort to using less sophisticated animals, such as mice, or would move aggressively forward with human trials. Regarding the latter, Stokes informed de Kruif that "we are greatly pleased with our results at the New Jersey State Colony" regarding the influenza studies. The institution had once been known as the New Jersey State Colony for the Feebleminded. The microbiologist-turned-journalist was apprised of an additional "experiment" at another "large New Jersey State Colony, which the state allowed [them] to use," but Stokes decided it would be best if that experiment, quite possibly a more sensitive one, was discussed when the men met in person.[9]

It was becoming increasingly common during the war years for the "feebleminded" to be viewed as that intermediate step in the research process between mice and "normal humans." Moreover, and tragically documented in the armed forces Committee on Medical Research files, state facilities for such special populations were considered the equivalent of frontline battlefield conditions. As one government document underscored, "In certain civilian institutions, where outbreaks of dysentery are not uncommon, opportunities have been furnished to observe the effect of the vaccines under approximately field conditions."[10] Doctors with access to such overcrowded and underfunded facilities found themselves sought after and much prized by the military, the academic community, and the private sector.

Dr. Stokes developed good relations with both the scientific community and the administrators who ran the large state institutions in southeastern Pennsylvania and southern New Jersey. Such relationships often made him the go-to guy for medical investigators seeking suitable "test material" for their experimental endeavors. One letter to Stokes, for example, was a request for access to mental institutions in Philadelphia or New Jersey. The writer needed subjects for trials of new encephalitis vaccines. The individual making this request was a doctor from the Yale University School of Medicine who was in need of "1,500 candidates, male or female, preferably between the ages of 17 and 40 years."[11]

By the last year of the war, New Haven, Princeton, San Francisco, and New York City had become hotbeds of experimental activity on hepatitis, poliomyelitis, and encephalitis. One of "the most important findings" of the experiments on hepatitis, for example, was that "the disease can be produced by feeding feces from a case of infectious hepatitis." To do so, however, required "human volunteers." These transmission studies "repeatedly reproduced" the feeding of feces laced with "infectious hepatitis" to test subjects. The key players in this drama, which "inoculated" dozens of subjects, included the Yale University School of Medicine, the Middletown and Norwich State Hospitals, and the Federal Correctional Institution of Danbury, Connecticut.[12]

In their effort to demonstrate that "gamma globulin will protect injected individuals against infectious hepatitis," "a girls' school in Providence, R.I. and a Catholic Institution for children in New Haven" were the subjects of study.[13]

Facilities for retarded and developmentally disabled children, such as the New Lisbon Developmental Center and the Skillman Center for Epileptics in New Jersey and the Pennhurst School for the Feebleminded in Pennsylvania became frequent travel destinations for doctors working under the auspices of the Army Epidemiological Board.[14] The institutional connections and travels of some investigators, such as Joseph Stokes Jr., could be easily tracked by their statements of appreciation at the conclusion of journal publications. For example, one thank-you note concerning a series of measles studies by Stokes, Elizabeth Maris, and several other doctors offered these words of appreciation:

> *Our thanks are due to all who have made these studies possible, particularly to Commissioner William J. Ellis and to Dr. Ellen C. Potter of the Department of Institutions and Agencies of the State of New Jersey; To Dr. Albert W. Pigott, Superintendent of Skillman; to Miss Ruth Jones, Superintendent at Mount Holly; to Mr. E. L. Johnstone, Superintendent at Woodbine; to Miss Marie S. Winokur, Superintendent of the Homewood School; to Drs. C. T. Jones and James Q. Atkinson, of New Lisbon; to Sister Mary Clare, Superintendent, and to Dr. Morris H. Schaeffer, physician of St. Vincent's Hospital; to Mr. E. Arthur Sweeny, Secretary of Welfare, State of Pennsylvania, and to Dr. James Dean, Superintendent at Pennhurst School. Also we wish to acknowledge the technical assistance of Mr. William P. Jambor in the preparation of the virus, and the assistance of Dr. Samuel X. Radbill in obtaining fresh clinical material.[15]*

Those receiving the authors' thanks had allowed children from their institutions to be incorporated in a series of measles experiments that included a challenge inoculation, in which "blood from children with active cases of measles" was injected into healthy children along with the forced "nasopharyngeal secretions from children with active cases of measles."[16] The several dozen children receiving the "challenge inoculations" resided in institutions such as New Lisbon, Pennhurst, Skillman, and the Homewood Orphanage in Philadelphia. Most of the trials, interestingly, were not initiated due to the urgency of war: They occurred *prior* to America's entry into World War II.

Not all investigative work took place in institutions. Army Epidemiological Board physicians could find themselves in some remote locations if the medical puzzle or scientific opportunity was deemed valuable enough. For example, in the summer of 1943, scores of young campers at Camp

Akiba in the Pocono Mountains had fallen seriously ill. The culprit was soon determined to be infectious hepatitis of epidemic proportions. The Army considered it their "good fortune to investigate an unusual and very interesting epidemic" in which "68 of 160 girls and approximately 40 of 170 boys had come down with the illness" that would continue to spiral into the hundreds.[17]

Believing the camp health crisis an "extremely interesting" event as well as an "opportunity to study the epidemiology of the disease in a closed group and to test the protective effect of the human immune serum globulin (gamma globulin) under natural or field conditions," army physicians traveled to the mountain region of northeast Pennsylvania and injected dozens of children with the serum, a blood derivative that Stokes had been trumpeting for years as a highly effective elixir for a host of infectious diseases.[18] As Stokes proudly informed the surgeon general's office, "The sudden cessation of all cases of jaundice in the injected group is the striking finding" and one that merits "larger field experiments."[19]

Medical investigators, however, did not need to tramp into the woods to wage war against infectious hepatitis and dysentery. They had the economic wherewithal and political clout to take over buildings for their studies, as they did at Stateville (Illinois) and Atlanta Federal Penitentiary for sophisticated malaria studies. Structures on college campuses could be seized as well, as happened in Philadelphia during the last months of the war. In their ongoing effort to conquer infectious hepatitis, doctors recruited fifteen conscientious objectors to be quartered for "60 days" in a "fraternity house on the campus of the University of Pennsylvania." During that time, different preparations were tested on the men after they were given the disease; some acted as a control group. Concerned about sick men living in close proximity to one another, not to mention in a congested college neighborhood, the government claimed that its "chief objective is to prevent transmission of the disease to other persons in the unit and to persons outside the unit."[20]

The hepatitis study on the Penn campus was deemed so important that the army brought in environmental engineers to spray the DKE house on South Thirty-ninth Street with a potent insecticide to ensure that mosquitoes did not interfere with the research during the coming summer months. However, the order to spray a stunning "240 gallons of DDT" on "the walls, ceilings, and window screens of the building" in "six treatments" leads one to suspect that the researchers may have been testing more than hepatitis remedies.[21]

More common than the isolation-type study on the Penn campus were the many "transmission" studies in which diseases were passed on to "volunteers" by "feeding feces preparations" to them. In fact, the Akiba strain of hepatitis from the Pocono Mountains children's camp incident became a source of future experimentation.

In addition to medical research emerging as a "well-coordinated, extensive, federally-funded team venture," the impact of the war years on human experimentation was not only "transforming" but also made for a "curious mixture of high-handedness and forethought" regarding the use of vulnerable humans as subjects for experimentation. In some areas, such as research into dysentery, malaria, and influenza, there was "a pervasive disregard of the rights of subjects—a willingness to experiment on the mentally retarded, mentally ill, prisoners, ward patients, soldiers, and medical students without concern for obtaining consent." However, as David Rothman argues, research in other realms, such as survival under hardship conditions and the sensitive area of sexually transmitted diseases, were more "formal and carefully considered."[22]

The discrepancy, he argues, was the result of decision makers' navigating a course of moral brinksmanship: "When they sensed the possibility of an adverse public reaction, they behaved cautiously." However, when they sensed that no one cared or was watching, they took liberties. Specifically, giving American citizens—even criminals behind bars—gonorrhea "might have raised a storm of protest from a variety of sources." The prospect of a controversial protocol appearing on the front page of a newspaper or in a courtroom was a worst-case scenario for researchers in general and for Committee on Medical Research directors in particular.[23]

Members of the profession were not going to allow that valuable resource to be cordoned off or circumscribed by an idealistic document designed for Nazi savages in a German courtroom. The Nuremberg Trial and Code were considered something of a historical anomaly; they had little application to the American medical experience and scientific arena. Doctors and researchers had grown comfortable using vulnerable populations for experimentation; they were not about to give up the freedom and luxury of mining a vast and relatively inexpensive resource.

And, in fact, they would not have to because no one seemed to care.

# FOUR

# IMPACT OF THE COLD WAR ON HUMAN EXPERIMENTATION

## *"There Weren't Any Guidelines as I Can Recall"*

KAREN ALVES REMEMBERS THE INCIDENT AS IF IT HAP-
pened yesterday, but the traumatic event was actually over a half century ago.
It was late May 1961, and she and her family were returning home from an
enjoyable day picking fruit and vegetables on a nearby farm when they heard
the phone ringing as they pulled into the driveway. Karen scampered out of
the car first, her two sisters trying to catch up, and her mother urging them to
answer the phone before the caller hung up.

When Karen picked up the phone and said hello, she heard an unfamiliar
woman's voice say, "I'm calling from Sonoma State Hospital. Is Rosmarie Dal
Molin there?" Karen was just ten years old at the time, and her heart sank.

She dropped the phone and bolted. "I ran and ran and ran," Karen
recalls. She instinctively knew the message was bad. "I knew my brother,
Mark, was dead. I don't know how I knew, but I knew. I ran to a secret hiding place and stayed there for hours. When I finally returned home long
after dark, my sisters were crying and my mother was locked in her room

talking in hushed tones to friends and relatives on the telephone. It was the worst day of my life."[1]

"My parents were told he had been sick. That he had been running a high fever and that led to a seizure and his death. But we were never told he had been sick, and he never had a seizure before. We didn't understand what happened. It was so unexpected."

Mark Dal Molin was born in 1955 with cerebral palsy.[2] "He couldn't talk or walk," says Karen, "but we loved him just the same. He communicated with his eyes, and we knew when he was happy or sad, when he wanted something from us or just wanted to play. He'd laugh and giggle and swing his arms and legs just as if he was playing with us."

The strain of raising a severely handicapped child, however, became an increasing burden on the Dal Molin family. Bill Dal Molin felt that Mark's care and needs were depriving the three girls of their mother's time, attention, and affection. Discussions about the situation between Karen's parents grew into debates and the debates into heated arguments. As she would subsequently learn, doctors were encouraging her father to have Mark committed to a state facility. They said it would be best for Mark and for the family.

"Mark was a vegetable in their eyes, but he *wasn't* a vegetable. He connected with us," says Karen. "We had fun with him, and he laughed with us when we played together." Bill Dal Molin, however, told his wife that Mark's care and increasing needs were hurting the family. It was time the state took care of him.

Three years old at the time, Mark was turned over to the state to be cared for at Sonoma State Hospital in northern California. That emotionally wrenching decision would haunt the Dal Molin family for decades. It would also leave lasting fissures between Karen and her parents and deep regrets for all.

Karen remembers coming home from school one day in late 1958 and finding the house eerily still. Mark had been taken away. "It profoundly affected me," recalls Karen. "I don't think I ever got over it."

Rosmarie Dal Molin had committed three-year-old Mark to Sonoma State Hospital, the largest institution for children in California. Over 3,500 children, many with severe birth defects, were housed there. Created in 1883 as the California Home for the Care and Training of Feeble-minded Children after two mothers crusaded for a suitable institution, by 1953 its name had been changed, and its mission emphasized medical treatment over training.[3]

Unlike many parents who gave up their children to state institutions and forgot about them, Rosmarie Dal Molin visited her son every Wednesday during the summer. Mark's sisters would visit on occasion as well, although Karen got the clear impression that institutional authorities were not enamored with the idea of young children visiting the hospital. "They thought we might be carrying the flu bug and wouldn't let us visit him. We'd play on the grounds outside like three monkeys doing all sorts of crazy things hoping to entertain him," recalls Karen. "We wanted to perform for him and make him laugh." Karen also recalls the "quiet trip home after every visit to Sonoma State. No one said a word."

According to Karen's sister Gail, their brother's death "pretty much blew the family apart. I believe that Dad did what he felt was best for the family. I know that is true. But the impact of it on each of us and the family was devastating."[4]

"The whole family was impacted by the decision to institutionalize Mark," says Karen. "Mark was a subject we didn't talk about. The family had problems. My parents became alienated and finally divorced. And I didn't know how deeply it had affected me until someone told me after observing my behavior, 'You're grieving. What are you grieving about?'"

Decades later, Karen was still haunted by her family's decision to give up her brother to the state. His sudden and unexpected death under what she considered suspicious circumstances only heightened her guilt. Sonoma State authorities told her parents that Mark had run a high fever and choked on whatever had lodged in his mouth. But Karen was dubious.

"I needed to know what happened, no matter what it was. I just needed to know and began looking into the case in 1993," she says. "I went to the recorder's office, but there was no death certificate. I talked to a clerk there, answered all his questions about when Mark died and where, and he thought for a moment and finally said, 'You might want to take a look at Sonoma State. Some strange things happened there during that time.'"

A news article she read describing how thousands of Americans—including children—had been used in radiation experiments spurred Karen's investigation. "The article," she said, "mentioned patients in state hospitals being used in medical experiments. I said to myself, 'That's it. That's what happened to Mark.'"

The search for documentation was slow and filled with bureaucratic roadblocks, organizational subterfuge, and long delays. Karen persisted; she was determined to find out what had happened to her brother.

About that time, President Bill Clinton ordered thousands of secret federal documents released detailing the government's involvement in human radiation experiments from World War II to the present day. The material only confirmed Karen's suspicions that her brother had been involved in dangerous medical experiments. In her pursuit of relevant documents, Karen discovered one study that verified that 1,100 cerebral palsy patients at Sonoma State had been incorporated in experiments between 1955 and 1960. One document mentioned her brother; he was also identified as "LPNI 8732."[5]

The July 1961 neuropathology report gave Mark's clinical diagnosis as "mental deficiency with athetoid quadriplegia due to hypoxia" and described various aspects of his brain from its weight to the "symmetrical atrophy of the thalamus and putamen." Karen was unacquainted with the medical jargon, but she was determined to learn every aspect of her brother's condition and the events leading to his death. She would become particularly familiar with a doctor's name at the bottom of the report: "N. Malamud, Chief, Neuropathology Service."[6]

The report confirmed a deep fear she had harbored—Mark's brain had been removed subsequent to his death and never returned to her brother's corpse. She recalled her mother saying at his cremation in the summer of 1961 that he didn't feel right when she touched him. The removal of his brain explained the mystery. His autopsy report, which was completed three days after his death, noted the probable cause as "a terminal type of aspiration pneumonia." It also stated that "the brain will be separately reported."[7]

Karen's hunt for additional documentation would continue. Much of what she learned showed that Mark and many other children at Sonoma State were inextricably linked to the research machinations of Dr. Nathan Malamud. A graduate of the University of California and its medical school, Malamud was a neuropathologist and director of the Langley Porter Clinic in San Francisco. He had become interested in the etiological origins of cerebral palsy and in 1953 filed an application with the National Institutes of Health

(NIH) "to investigate the various causative factors operating in the production of cerebral palsy."[8]

In his application, Malamud stated that "biochemical, pneumoencephalographic, angiographic, and electroencephalographic" diagnostic tools would be used and that "a large sample of patients with cerebral palsy from Sonoma State Home [would] be studied intensively." Karen shuddered at the thought of young children being put through diagnostic tests such as pneumoencephalograms, extremely painful procedures in which air is injected into the spinal cord and through the brain in coordination with a series of X-rays. "Imagine puncturing someone's spinal cord, drawing fluid out and putting a foreign substance in there like gas," she says. "When they trap air in your body, you're in pain, excruciating pain, for days."

Another feature of Malamud's proposal not only allowed for such children to be "thoroughly studied during their lives" but added the prospect of a "greater number of individuals coming to postmortem examination." Malamud went on to estimate that during "a period of five years a potential group of 1200 cerebral palsied individuals would be studied historically, genetically, and clinically. Based on our experience of the past six years," argued the doctor, "it might be expected that approximately 150 of these cases would come to post-mortem examination."

Malamud's proposals filed with the NIH during the 1950s disclose that "toxic solvents" and "noxious gases" would not be employed in any of the research endeavors. The same could not be said for "radiation," however, which was clearly checked off on application forms as an integral aspect of the study. And it is a certainty, at least according to Karen Alves, that her brother was negatively impacted, if not killed, by radiation poisoning.

"Dr. Malamud injected radioactive material into Mark's spinal cord," Karen declares. "It wasn't designed to help him; it was just a tool for researchers to learn more about cerebral palsy. He and the other children were just used as guinea pigs to attain more knowledge about the disease."

Malamud and his team of researchers would write several journal articles about their findings. One 1964 publication summarized their work between 1955 and 1960: "During this time, a total of 4,843 patients were screened of whom 1,184 or 24.5 per cent, were classified as having C.P. Of these, 508 cases were selected as the final group. . . . Approximately 20 per cent of the selected group have thus far come to autopsy and this preliminary report is based on an analysis of 68 of consecutively autopsied cases."[9]

The loss of her brother and her long investigation into what actually occurred at Sonoma State and the Langley Porter Institute have dramatically impacted Karen's opinion of science and the doctors who practice it. "The researchers were a very cavalier group," she says. "They were arrogant and had an ability to disregard those who were suffering. Just like the victims of eugenics years earlier, people like my brother had no say in what was being done to them. They had no voice. And the doctors' need-to-know mentality drove them to do things that they shouldn't have. They were able to shield themselves from ethics, conscience, and compassion."

Her study of the era has also colored her view of the politics that produced and encouraged such risky scientific ventures. "The fear of an all-out nuclear war," argues Karen, "drove the research community to look deeper into the effects of radiation and the treatment of radiation injury. I believe Malamud was doing cerebral palsy studies for the Manhattan Project. They were into gathering data, and national security outweighed concern for the public or compassion for patients and research subjects. . . . Back then doctors were seen as gods; you didn't challenge them."[10]

Despite the passage of decades, the many emotionally charged family discussions, and the years of historical research, Karen continues her search for information about Malamud's work, the relationship between Sonoma State and Langley Porter, and the whereabouts of her brother's brain. "They took Mark's brain without ever asking my parents' permission," says Karen. "Malamud's obituary said he had one of the largest brain collections in America; hundreds, maybe thousands of children's brains. I'll look under every stone until there are no stones left."

There were many like Mark Dal Molin, children with various physical maladies or mental challenges who were sequestered from society only to be further exploited in the name of science. Like Karen Alves, many families are becoming aware of past medical practices and learning how loved ones were conscripted into nontherapeutic experimental research during the Cold War.

Ironically, many of these recently discovered instances of medical abuse occurred after we had proudly trumpeted our abiding devotion to the highest standards of research ethics at Nuremberg. The unilateral decision by the United States to prosecute the Nazi doctors for their brutal experiments and junk science was meant, in part, to showcase America's principled stand regarding human research. Unfortunately, America's track record, then and now, is less than exemplary, and our post-Nuremberg record on research abuse would grow more problematic with each passing year.

**IN AN EFFORT TO FURTHER** the medical community's knowledge of hepatitis, an experiment was initiated in the early days of the Cold War whereby five volunteers were inoculated with the Akiba strain of the virus. When all five volunteers showed "suspicious findings," including "hepatic disturbance," it was decided to do a follow-up experiment using additional subjects. The protocol called for nine volunteers ages ten to fifteen in three separate groups to receive tainted Akiba stool serum. Not surprisingly, "mild clinical illnesses and definitely positive laboratory results were noted in one or more volunteers of each group."[11]

The "volunteers" in these experiments were children at the Pennhurst School in southeastern Pennsylvania, an institution built in the early 1900s to house the "feebleminded." One child, a ten-year-old boy, was now suffering an assortment of ills, including malaise, nausea, vomiting, constipation, and liver inflammation, not to mention an assortment of other maladies.

The other volunteers, as they were euphemistically called, were given hepatitis-laced feces to eat and emerged with many similar symptoms.[12] Seven of the nine children were diagnosed with either "liver tenderness" or "abnormal liver function" after the experiment.[13] There is no mention that parental permission for the children's participation had been secured.

The Pennhurst transmission study was completed in August 1947. The date is significant. This research was occurring at the very same time that an American tribunal in Germany was prosecuting Nazi doctors for violating accepted standards of research. The timeline was also consistent with American-inspired experiments in Guatemala that were transmitting various sexually transmitted diseases to soldiers, prison inmates, and children.[14]

The vast majority of Americans, however, were ignorant of both the hypocrisy that was at play and the many unethical and dangerous experiments that were occurring in the research arena. Institutionalized children would become increasingly popular as test subjects. Joseph Stokes Jr. of the Children's Hospital of Philadelphia and the University of Pennsylvania School of Medicine had long-established relationships with institutional officials and could usually locate test subjects easily. His research on hepatitis after the war would continue with Pennhurst providing a seemingly endless supply of "research material." Transmission studies whereby "pooled feces" suspensions were given to children through their allotment of milk was just one of many "challenge" studies and jaundice experiments at the Pennsylvania institution during the summer of 1947.[15]

Physicians' reports and personal communications from this period disclose a rather seamless transition between wartime and postwar scientific endeavors. Human experimentation continued unabated and would soon increase in scope with an ever-larger number of vaccines and products under investigation, research sites being utilized, and subjects participating in clinical trials. In fact, many medical historians, have aptly referred to the era they were entering as the Golden or "Gilded Age of Research" in America.[16]

"Many scientists and political leaders," according to medical historian David J. Rothman, found it unthinkable that so critical an activity as public health and scientific research would be "permitted to regress to prewar conditions of limited and haphazard support by private foundations and universities."[17] Collective effort and solid financial support had produced unprecedented medical triumphs during the war years. Longtime threats such as smallpox, typhoid, tetanus, yellow fever, and various infectious diseases had finally been conquered, and it was thought that medical science was at the threshold of even greater discoveries. It would be foolish for government support of science to be withdrawn when such important breakthroughs beckoned.

It was a position that was hard to argue with. "Miracle drugs" and those who discovered them had won the day. Penicillin alone was widely hailed as a lifesaving wonder drug that had tamed if not wiped out numerous maladies and sparked the imagination of scientists and nonscientists alike. Potential elixirs that might cure polio, cancer, and other dread diseases might be next. Alexander Fleming, the British scientist who had documented the medicinal powers of a simple mold and transformed it into penicillin during World War II, became an international celebrity and a role model for legions of aspiring microbe hunters around the globe.

With both politicians and the public in agreement that government investment in research should continue, it was quickly decided that the National Institutes of Health would replace the wartime Committee on Medical Research as the vehicle for conducting an all-out campaign against disease. Over the coming years, NIH would be the recipient of a tremendous infusion of cash. During the last year of the war, it received less than $750,000. Over the next ten years, that sum would grow to $36 million; it was well over ten times that by 1965; and by 1970, it reached $1.5 billion with some 11,000 grants being awarded to enterprising researchers and institutions. A third of those grants required some form of human experimentation; infants and children were often part of them.[18]

Not surprisingly, greater value was placed on scientific triumphs and the attainment of knowledge than on the welfare of those incorporated as test subjects in the research. As Rothman states of the operations of NIH's own Clinical Research Center, "the hallmark of the investigator-subject relationship was its casualness, with disclosure of risks and benefits, side effects and possible complications, even basic information on what procedures would be performed, left completely to the individual investigator."[19] This laissez-faire approach to scientific investigation would dominate the research arena from the end of the war until the mid-1970s.

The Nuremberg Code, it would appear, had little to no impact on the American research community's growing prominence and operations. It was almost as if the trial of Nazi doctors and the subsequent establishment of a code of research conduct had not occurred or just pertained to those who practiced medicine like Nazi savages. The result was the same; the event had little consequence for American medical practitioners. Granted, the American Medical Association (AMA) mentioned the Nuremberg Trial in its journal, but one would be hard pressed to say the organization stressed its importance. One rather brief account of the proceedings published in November 1947 under the heading "The Nuremberg Trial against German Physicians" listed both Nazi medical transgressions and the new code's principles. A careful reading of the code, however, shows that it was already being deemphasized, especially those clauses journal editors might have construed as overly restrictive.[20]

In the article, the code's first principle, for instance, was actually 180 words short of its original formulation. The key voluntary consent provision was mentioned, but critical passages concerning the test subject's capacity to exercise free power of choice were expurgated. So also would be exclusionary factors such as force, fraud, deceit, constraint, and coercion, not to mention the subject's ability to make an enlightened decision. Protective mechanisms designed to safeguard a subject's health were thought unimportant. That the bulk of the first principle's contents were deleted mere months after the code was created was an omen of what was to come.

The AMA wasn't alone in deemphasizing important ethically related subjects; medical schools were equally complicit in underplaying the importance of such issues. As the Oral History Project of the Advisory Committee on Human Radiation Experiments discovered during historical research for its final report, none of the physicians interviewed recalled receiving formal training in medical ethics during their medical training. In addition, few recalled any formal institutional review of research protocols being required prior to the early 1960s.[21]

Our own research confirms the Oral History Project's findings that "the Nuremberg Code had little salience for American biomedical researchers. Few recalled any discussion relating to the Code at the time of its issuance, although they certainly remember the atrocities of the Nazi doctors and the war crimes trials in Nuremberg."

Numerous interviews we conducted with renowned medical researchers who received their medical training in the 1950s and 1960s corroborated the absence of the Nuremberg Code and any semblance of serious ethics instruction in their course work. "I was unaware of the Nuremberg Code and its code of conduct," admitted Dr. Chester Southam. "I had no awareness at all."[22] "There was no training in medical ethics at the time," admitted Dr. A. Bernard Ackerman. "No one ever brought up the Nuremberg Code."[23]

Despite the lack of formal course work on ethics-related subjects, researchers weren't oblivious to moral and ethical concerns. The Nuremberg Code may have been perceived as a distant and idealistic construct, but individual physician-investigators on this side of the Atlantic faced an array of ethical conundrums. They frequently debated among themselves a cross section of ethical puzzles and moral quandaries running the gamut from who could be used as an experimental subject and whether remuneration for volunteers was in order, to the challenge of keeping controversial research from the public and the "yellow press."

An examination of the activities of one busy and strategically situated physician-investigator illuminates the postwar moral and ethical landscape regarding human research and the difficulties inherent in plotting an acceptable course between unfettered experimentation and the newly formulated but overly restrictive Nuremberg Code. Dr. Joseph Stokes Jr. would make his mark as one of America's leading viral and vaccine specialists during the middle decades of the twentieth century. An ardent champion of human research as the best way to relieve and prevent childhood diseases, Stokes was no Nazi and can in no way be compared to the physicians consigned to the gallows for their horrendous crimes. But he was willing to use vulnerable populations as volunteers where others advised greater caution and a less vigorous path to new vaccines. Stokes's correspondence with research colleagues during the middle decades of the twentieth century reveals the complexity and subtleties of high-risk, high-reward medical research in postwar America.

Though a staunch advocate of medical research, Stokes was not blind to the possibility of unexpected consequences or to the sacrifices of test subjects. He often expressed his concern for prison inmates involved in research. He thought their involvement in clinical trials important and necessary, but he did voice his concern about those negatively impacted by experiments gone wrong. He thought that reimbursing volunteer prisoners who might be disabled as a result of the studies was only just, and he knew that the "viral agents" under investigation were potent enough to cause lasting injury. As he wrote in one letter, "Our viruses have not apparently been very virulent strains and we do not feel particularly worried about permanent disability; there is always, however, a bare chance."[24] He noted in another letter that such research "might raise a point of ethics," and that reasonable compensation was probably in order.[25]

Unremarkably, self-interest usually won the day. Colleagues of Stokes urged him "to protect himself . . . by means of the usual waiver" in order "to release the experimenter from any liability." Though the doctors themselves didn't know whether "such a waiver would really be of value in case of a suit or death or disability at a later date," it was generally considered a good idea.[26] The original issue—payment for injured test subjects—became of secondary interest; the greater issue of concern was physician liability.

Adding his own sentiments on the moral and legal issues confronting them, the director of the Commission on Virus and Rickettsial Diseases of the Army Epidemiological Board (on which Stokes served) agreed that "ethics" was very much at issue, but he offered caution since they were navigating through uncharted waters. Wrote Dr. John R. Paul:

At this stage in the world situation one should proceed cautiously, until standards are set up by what ever body is in 'authority.' I am not sure just what the rules are but I understand that Dr. Ivy at the University of Illinois has been on some type of vigilance committee which has laid down certain principles about volunteers in order to protect this country from the criticisms brought up in Germany during the Nurnberg [*sic*] trials. The Russians in Japan have also accused U.S. scientists of experimenting on humans. During the war we more or less made our own policies on this, but I am not sure that that is possible today and if there are to be official policies, I believe, we have to know them before any official statements could be made.[27]

Despite his interest in the economic welfare of injured prisoners, Stokes had no such concerns regarding his use of children in studies across the country. In defense of his work, he repeatedly argued that no research was begun that would not be of benefit to the individual chosen for the investigation. From Stokes's point of view, "a planned exposure to a known MILD virus under nursing care during incubation" was superior from a health perspective to "an unplanned exposure to possibly a more virulent virus strain at a more dangerous age period after puberty."[28] In coming years, Stokes would repeatedly fall back on this line of argument as critics questioned his use of institutionalized children as test subjects.

Such a rationale would appear rather lame in light of some of his research. Consider his work with an Illinois physician who was doing epidemiologic studies of infectious hepatitis in children at a Chicago orphanage. The project, which ran from July 1951 to July 1952, set up a "special ward where non-immunes [could] be exposed to active cases" and thereby test Stokes's theory regarding the disease-prevention attributes of gamma globulin. There is no mention of parental permission, but it can be assumed that the facility's superintendent had no objections to exposing the young wards in his care to a dangerous liver disease if it would foster the advance of science. Further control studies were to be carried out on "infants with a variety of infections" at another hospital. When close proximity between healthy and sick children seemed lacking, "stool preparations obtained from 2 children in the St. Vincent's [Chicago] orphanage study who were suspected of being chronic carriers for 6 and 16 months respectively [were] administered orally" to research subjects, thereby insuring "fecal–oral spread" of the disease.[29] Research physicians frequently took such liberties with their test subjects. No one objected except perhaps the orphans, and their wishes may not have counted.

Stokes would be a key player in the government's establishment of medical policy on the use of test subjects. As the chairman of the Sub-Committee on the Allocation of Volunteers on the Commission on Liver Disease of the Armed Forces Epidemiological Board (AFEB), the successor to the Army Epidemiological Board, Stokes was at the heart of the many debates and decisions on who and what populations could be used in human research. His sub-committee colleagues included some important names in American research.[30]

The army, which had targeted nearly three dozen diseases requiring further investigation by AFEB physicians and divided them into categories of military importance, understood the need for human subjects as test material.[31] "Without volunteers, however," stated Stokes in one missive to an AFEB colleague, "the obstacles at times seem insuperable and I hope . . . every effort to obtain them" is made. He didn't believe they could expect much success without them.[32]

Stokes was quite assured of his stand on the use of volunteers, including children, in his research initiatives, but other equally accomplished scientists occasionally voiced their doubts. Communications between committee members illuminate the personal and political fallout from their disagreements. When a colleague upheld a New York City "Sanitary Code" precluding the use of children in one study, Stokes expressed his dismay and notified his superiors that such rules should be fought.[33] He argued that it was "a moral obligation to lift" the ban. He said the opposition to his experiments on children was misguided and added, "I have exposed my own children and aided my physician son in the exposure of his first child as the opportunity arose."

Undaunted by those advocating a go-slow approach, Stokes believed, "There is no reason to beat a retreat on these problems but to face them positively and at the same time with the greatest deference to the value of human life and welfare."[34]

Stokes wasn't just concerned about his own research endeavors; he was also irritated when experiments of other investigators were blocked. He remained steadfast in his opposition to the imposition of overly restrictive guidelines and regulations on the use of children in medical research. When Dr. L. Emmett Holt, for example, had his request for funds for a study of infants' amino acid requirements rejected, Stokes was miffed. "Here again," he wrote to a colleague, "it seems clear that a single member of the original Committee, which considered Holt's application, quite unreasonably injected ethical considerations into a matter which is of the utmost importance to human nutrition . . . and can be of no harm to the infants."[35] Stokes went on to say, "Once the Nuremberg specter (and I am in agreement that the Nazi work was inexcusable and despicable) is raised, it appears at times that all rational approaches to such matters are obscured."

Stokes saw the Nuremberg shadow as an impediment to medical research and equated it to ethical "bullying," a tactic he thought particularly prevalent in New York City. Since Holt's investigations in Galveston were similar to Stokes's, he had hopes that the "medical bullying which our major metropolis seems to occasionally nurture" would not occur in Texas and that Holt's application would receive the dispassionate consideration it deserved.

Interestingly, Stokes was a Quaker who saw the benefit of conscientious objectors offering themselves to science instead of succumbing to bullets on the battlefield. Recognizing the continuing need for volunteer subjects after the war, he advocated a continuation of the program and encouraged the American Friends Service Committee to consider "using some of the . . . available 18 year old group on a draft deferred basis as experimental subjects."[36] Stokes was not alone in his belief that "pacifists" could best serve their country and science by becoming test subjects.

It was estimated that over 1,000 men had participated in such work, and the presence of a group of healthy, cooperative subjects was a "stimulus to scientific investigation and a quite unusual opportunity for the investigator."[37] Advocates of continuing the program believed that the experiments were a "service to mankind as a whole, with potentialities for the saving of lives which can hardly be over-estimated." But there was also a downside. Some test subjects developed psychiatric problems, and doctors increasingly felt the need to get legal waivers from participants and permission slips from parents.

The Quakers had every right to be cautious when allowing their young men to participate in human research; experiments were dangerous, and some had proved fatal. Members of Stokes's volunteers allocation subcommittee were unnerved after three prisoners died and a fourth became very ill. But any hiatus would be fleeting; there were many more reasons to continue the research rather than to terminate it. One reason was that those enduring the health risks were some of the least valued members of society: incarcerated criminals, psychiatric patients, and the developmentally challenged. Many of the latter group were children—a population without political influence or social capital. As one observer noted, "If it is true that the progress of medicine has been over a mountain of corpses, one objects to its being one's own corpse. If it is also true that in medicine 'nothing risked, nothing gained,' one prefers to have the gain to humanity made at the expense of somebody else."[38]

The "somebody else" was often an impaired child at a place like Vineland, Pennhurst, Fernald, and Sonoma State—hardly the institutions with the same social cachet as schools for so-called normal children. Besides, as Dr. Colin M. MacLeod, the president of the AFEB, informed a New York State Health Commissioner, such studies had "military significance." It was important from a national defense perspective that the fight against epidemic hepatitis and serum hepatitis be successful. Despite their best effort, MacLeod would point out, "it had not been possible to transmit" infectious agents to experimental animals. The upshot was clear: The use of human volunteers was obligatory if diseases were to be conquered.[39] He admitted to being concerned about the risks involved in human experimentation, but the advance of science was critically important to solving "crucial military problems." And "access to human volunteers" was essential to the system's success.[40]

Stokes had no doubts about that; he understood the value of test subjects and was sensitive to attitudes or initiatives that limited their availability. He drafted a confidential memo on the "question of Ethical Responsibility in the Exposure of Human Beings to certain Infectious Agents." The February 1953 document focused on the medical risks of exposure to epidemic disease and also on the criticism that "conscious exposure of volunteers or children is morally or ethically wrong."[41]

Stokes believed researchers should not adopt "a defensive attitude" or feel "vulnerable on moral or ethical grounds" owing to the occasional criticism they received. Despite the occasional mistake, they had every reason to stand by their work and continue their important mission with "candor and forcefulness."

What may be most noteworthy about the six-page, 1,000-plus-word memorandum is not what is included in the document but what is missing: Stokes made no reference to the Nuremburg Code. Clearly, the code had no import or relevance for Stokes. Yet he was a major figure in medical research at the time and was central to shaping AFEB and government policy on the use and allocation of test subjects for military- and university-sponsored research during the Cold War.

The use of prisoners in research was a comparatively easy call for Stokes; other vulnerable populations were harder to justify. As he admitted in his memo, "a special problem arises in the case of children."[42] Stokes defended the practice by arguing that whenever children were used in studies, permission had been obtained from their parents or guardians, but he conceded that "the notion lingered that the children are being used involuntarily and that this is somehow wrong." Though he thought this aspect "the nub of much of the present criticism," he believed that it had already been addressed through commonly accepted practices such as immunization programs for children. In other words, children don't give their "consent to inoculation against diphtheria, tetanus, and whooping cough," but the effort to minimize "loss of life, illness, or permanent injury from these diseases has been felt far to outweigh any disadvantage to a child's rights."

Was Stokes confusing therapeutic measures with experimentation? That would seem too simplistic an answer for a man of his experience and sophistication. It is clear, however, that he—and no doubt many of his colleagues doing viral research—believed that knowledgeable, well-trained infectious disease specialists knew what was best for both the patient and the subject. The real problem, according to Stokes, was "one of medical advisability and not ethics." Stokes was adamant on the subject; he recognized that "differences of opinion exist, but the problem is not one of ethics." In his mind, protective codes that tied the hands of researchers and restrictive regulations "in the name of ethics were actually self-defeating." The formal establishment of such a "concept of ethics," argued Stokes, "is not only mistaken but harmful." In the end, he repeatedly urged his colleagues to move forward aggressively. It is obvious that medicine, he advised them, "can be no more static than in any other field of investigation."[43]

Though many of the top researchers at the time shared Stokes's opinions on these questions, not all were as strident in championing the cause. Most desired an unfettered research arena but also wanted to leave public debate to others. Dr. Roderick Murray of the National Microbiological Institute, for instance, cautioned Stokes that little would be gained by bringing the issue of volunteer studies to public attention. "By any active action," he argued, "we might be put in the position of protesting too much."[44] In his missive to Stokes regarding the uphill climb lying ahead of anyone attempting to explain the nobility and importance of their mission, Murray cited the "recent speech by the Pope on the subject of medical experimentation."

The papal reference was in regard to a September 1952 speech by Pope Pius XII titled "The Moral Limits of Medical Research and Treatment." In it, the pontiff paid homage to science, "the bold spirit of research," and the researcher's zeal to follow "newly discovered roads." But he also identified some cautionary signposts, moral landmarks he thought men of science occasionally chose to disregard.

The quest for new knowledge is important, argued the pontiff, but that didn't legitimize every method designed to attain it, especially if it could not be accomplished without injuring the rights of others or violating some moral rule of absolute value. Science is not the highest value that all other values should be subordinated.

Though the speech did not refer to the Nuremberg Code, the pontiff's intention was clear; there are "many values superior to scientific interest."[45]

The speech, which would emphasize "*bonum commune*," or interests of the community, over the "interests of science," was a papal warning shot across the bow of a rapidly growing and increasingly robust research establishment. But as in the case of the Nuremberg Code, no one of substance and no organization with clout were willing to support the cause and campaign for reform. Most did not hear calls for a more ethical research culture; those who did tended to forget about them.

A high-powered, financially rewarding, and celebrity-oriented template for success had taken hold. Martin Arrowsmith was no longer an outlier; Paul de Kruif's solitary and poorly funded microbe hunters had evolved into a much-admired and potent professional force. Periodic calls for greater restraint and concern for subject safety by a few jurists in Germany or a religious figure in Rome were not going to stem the tide. Change would eventually come, but it was still decades off.

**MEDICAL RESEARCH DURING THE COLD WAR** years would become more sophisticated, lucrative, and imposing. Research using those behind bars, for example, witnessed a geometric increase, and researchers like Dr. Stokes were not shy about defending the practice. As he wrote in one memo, "It cannot be emphasized too strongly that such work with volunteers in prisons conducted under the Armed Forces Epidemiological Board has no element of compulsion. The objectives of the work, its exact nature, and the possible dangers entailed are fully explained to the prisoners." Moreover, "no promises of earlier parole or release from prison or other inducements" were tendered to procure volunteers.[46]

Such a belief would appear to be willfully ignorant of the fact that prison "volunteers" were behind bars, and constraint, coercion, duress—key elements of imprisonment—were specifically identified in the first principle of the Nuremberg Code as disqualifiers for participation in human subject research. Dr. Werner Leibbrand, a German physician and prosecution witness at Nuremberg, vigorously argued that prisoners should be precluded from participating in human research. Stokes, and most of his research associates, didn't see it that way. They did not view confinement behind concrete walls and steel bars as coercion. And despite his claim that no inducements were offered the prisoners, the twentieth-century history of such operations is one of inmates expecting and receiving something in the way of remuneration, such as money, early release, or better living conditions in the institution.[47]

Prison research was becoming big business. And prisoners kept in the dark about the dangers of such work, but too quick to see the benefits, "volunteered" as human guinea pigs in impressive numbers. So many, in fact, that researchers had the luxury of winnowing out those who might not possess the ideal public profile of a test subject.

For example, in December 1952 both the AFEB and the AMA adopted a resolution that expressed "its disapproval of the participation in scientific experiments of persons convicted of murder, rape, arson, kidnapping, treason or other heinous crimes."[48] Apparently, becoming a test subject behind bars had accrued some degree of status and prestige; it was now a privilege that researchers didn't want sullied by violent felons.

Stokes and his colleagues on the AFEB were very conscious of the explosive nature of their studies, the constant threat of negative publicity, and the importance of what the public learned about their research. They usually had other board members review statements, journal articles, and other items regarding the use of "retarded children" and "psychotic criminals" in studies. Stokes was constantly on the alert for "yellow press publicity." He believed there were some in the print media who rarely missed an "opportunity for criticism."[49]

Stokes was not alone in his concern. When journalists expressed interest in doing a story on their research endeavors, it precipitated collective unease. In early 1953, for example, the AFEB president received a letter from a *Washington Post* reporter requesting information regarding the use of prisoners in medical research. Though the reporter, Nate Hazeltine, said that his account would be factual and objective regarding the use of prisoners as test subjects, alarms still went off at AFEB headquarters.[50] Members of the AFEB voiced their concern about the prospect of an unflattering newspaper story; a potential "hornets [*sic*] nest," in the words of one army physician. They did their best to dissuade the reporter from writing a story.

If the prospect of a news story about prison research unnerved them, one can imagine their angst if a reporter had gotten wind of their experiments using "defective children."

The correspondence between doctors and their research associates illuminates some of the occasionally deceptive practices and cavalier decision making involved in the endless search for "test material," the identity of the actual study sponsors, and the health risks involved. A prime example of such subterfuge occurred in the summer of 1954 when Harry Von Bulow, the superintendent of a South Jersey institution formerly known as the Woodbine Colony for Feeble Minded Males, wrote Stokes about some growing concerns with "our little project." The project was an experiment to test a new poliomyelitis vaccine, "an active virus vaccine by the natural route, the mouth."[51]

Von Bulow and the Woodbine board "were not too happy" that the project was initially broached and agreed to over the telephone and that key concerns about children's safety never received a proper hearing. A further concern was that information about the oral vaccine experiment was misleading and might "destroy any confidence" the institution had established with parents over the years.[52] As letters demonstrate, researchers and administrators held different notions on just how much information parents were entitled to receive regarding their child's participation in a potentially dangerous polio study. In one revealing admission, Von Bulow wrote Stokes, "I am sure you can understand my administrative anxiety and my desire to inform parents as completely as we can without actually telling them what is taking place."

Von Bulow was no administrative novice. In order to win support for the project from the Woodbine board, he requested that Stokes write a letter to them that said "there were no risks involved either to the children concerned with the project or with the others in our Colony."[53]

In another revealing missive concerning the same piece of research, Stokes notified a Lederle Laboratories division head that the proposed letter to parents concerning the experiment had been altered to omit "the name of Lederle" substituting "the supervision by University of Pennsylvania physicians" since they "did not wish the parents to obtain the impression that this is primarily a commercial venture."[54]

Not surprisingly, the Lederle Corporation was "heartily in accord" with Stokes's actions. The viral research director wrote him a thank-you note stating: "We both feel it a course of wisdom to not mention the name of that reputable industrial company which will prepare the vaccine. I think that it is only smart because of the legal entanglements that might ensue should our participation in this program become general knowledge."[55] Once again, the less parents knew about what was going to be done to their children, the better.

With all the research projects that were under way, it is no wonder that concerned parents occasionally became skeptical and wrote to an institution's superintendent or medical director to learn if their child had become an experimental lab rat. On at least one occasion, Stokes felt obligated to inform an outraged mother, "I am writing to let you know that the children at Woodbine are not being used as guinea pigs."[56] In letters of this type, he would go on to explain the purpose of such studies and attempt to mollify strained relations, but some parents remained dubious.

FOR THE REST OF THE 1950S AND BEYOND, audacious medical researchers would proceed with the fervor of evangelists. Whether motivated by altruism, career advancement, or the attraction of old-fashioned fame and fortune, all manner of scientific quests, intellectual puzzles, and baffling health concerns became grist for the research mill. Medicine and medical research during the postwar era witnessed phenomenal growth. The number of people working in some field of medicine grew exponentially, climbing from 1.2 million people in 1950 to nearly 4 million in 1970. Health care expenditures climbed even higher: from $12.7 billion to $71.6 billion during that same period.[57] As its esteem and respect grew, medical science was accorded unprecedented approval by an appreciative public, and its role in protecting the nation became a widely accepted fact.

As Harvard medical sociologist Paul Starr has written, "medical science epitomized the postwar vision of progress. . . . All could celebrate the value of medical progress," take pride in the creation of "the latest wonder drugs," and be thankful that "life was getting better." *Time* magazine was now devoting a page in each issue to medicine, where Americans could learn of newly discovered "wonder drugs" and miracle treatments from contemporary microbe hunters and sophisticated laboratories.[58]

And during this period—an era ominously known as the Cold War— science assumed a symbolic and practical function in maintaining America's position as the unrivaled leader of the free world. It was not some honorary role requiring little concern and even less work; the threat from the Soviet Union was real, and its scary portent leached into every aspect of American life. As Federal Bureau of Investigation chief J. Edgar Hoover repeatedly counseled his countrymen, every American home was making a sacrifice in order to keep our defenses strong against the world-wide advance of Communism.[59]

Though he may have been one of the more ardent and outspoken Cold War warriors calling for constant vigilance and preparedness, Hoover was not alone in his fear of the growing "Communist menace." Americans had gradually come to view our World War II ally as our greatest postwar threat. A quick succession of events underscored the looming conflict. The Soviet blockade of Berlin in 1948, the Soviets detonating their own atom bomb a year later, and the start of the Korean War in 1950 all confirmed Winston Churchill's specter of an "iron curtain" separating the free world from the enslaved and contributed to the increasing prospect of violent engagement. Adding to the dark diplomatic developments was the equally ominous domestic threat of internal subversion. Names such as Alger Hiss, Whittaker Chambers, Elizabeth Bentley, Harry Gold, Klaus Fuchs, and Ethel and Julius Rosenberg among a host of others confirmed the Soviet Union's evil intent. Rarely a day passed when newspapers did not carry a story dealing with House Un-American Activities Committee (HUAC) hearings, new charges of subversion and espionage being hurled by Senator Joseph McCarthy, and a rising number of Americans building underground bomb shelters while their children practiced duck-and-cover drills in schools throughout the land. By January 1951, the Federal Civil Defense Administration (FCDA) had already published manuals about the dangers of biological warfare; they warned of agents such as botulism, plague, smallpox, cholera, and anthrax that might be sprayed over cities in deadly aerosols or put into supplies of food and water.[60]

The outbreak of a new war between the United States and the Soviet Union wasn't the creation of a novelist or Hollywood screenwriter. Many Americans anticipated a cataclysm that was fostered by a quasi-religious belief that "the Communists were Satan's army on earth."[61]

And the feeling was mutual. Soviet leaders uniformly agreed that the United States—a bulwark of capitalist enslavement—was "the USSR's main adversary."[62] As Richard Rhodes aptly said of the era, for serious men and women on both sides of the Iron Curtain, the encroaching Cold War cast an apocalyptic specter.[63]

Medical researchers were not oblivious to the ever-worsening political climate. "Although it was not a shooting war," Dr. James Ketchum, a highly trained and competent psychiatrist and psycho-pharmacologist for the US Army during the 1950s and 1960s, has written, "the stakes were every bit as high as in the World War that had only recently ended. Most ominously, each nation had the ability to launch megaton nuclear missiles sufficient in number to annihilate the other. Popular novels and films trafficked in visions of Armageddon and apocalypse. The phrase 'mutual assured destruction' became linguistic currency among journalists and commentators."[64]

The upshot, according to one comprehensive government study of the era, was the "likelihood that atomic bombs would be used again in war, and that American civilians as well as soldiers would be targets, [which] meant that the country had to know as much as it could, as quickly as it could, about the effects of radiation and the treatment of radiation injury."[65] The search for knowledge and answers to critical military and public health queries required scientists and medical personnel to once again confront questions of risk and what actions need be taken to protect Americans. Physicians who had been educated during the heyday of the eugenics movement and went on to perfect their medical skills during World War II when ethical concerns were subordinated to national security interests were once again finding their "commitment to prevent disease and heal" subverted by the government's immediate needs. It would also prove an "opportunity for gathering data."[66] As one Central Intelligence Agency (CIA) physician informed a classroom of recruits in the 1960s, "Our guiding light is not the Hippocratic oath, but the victory of freedom."[67]

"There was the definite feeling of the threat of war with the Soviets," recalls Ketchum.[68] A graduate of Dartmouth and Cornell Medical School, Ketchum makes clear that the threat wasn't created out of boredom or the lust for war; concrete events and intelligence signaled potential strife. He says it was clear that the United States had fallen behind the Soviets with regard to its chemical warfare capability and the immediate need to come up to speed in preparation for some future enemy attack. It was equally clear to Ketchum and his military superiors that research scientists "could not fulfill this mission by animal experiments alone."[69] Human test subjects would be needed; fortunately, most experienced researchers already knew where to look for volunteers.

Physicians working for the military would prove particularly adept at testing innovative theories on isolated test populations while maintaining a strict code of secrecy. Spurred on by patriotic fervor, the prospect of discovering medical breakthroughs, and provided with more than adequate funding, doctors on the government payroll blazed a trail of vigorous experimentation that explored everything from truth serums and incapacitants to mind control. Once again, concrete results would be the coin of the realm, and abiding by codes of ethics was a luxury they couldn't afford. Many a historian as well as numerous physicians would point to the Cold War atmosphere as the excuse for the numerous code violations, medical excesses, and instances of potential, if not actual, criminal behavior committed during the postwar era.

One of Ketchum's army medical colleagues, Dr. Enoch Callaway, recalls "the very lax research atmosphere" during the postwar years. There were "no codes stressed back then or any mention of the Nuremberg Code." Callaway graduated from Columbia Medical School in 1947 and subsequently spent many years exploring offensive and defensive weapons including nerve gas. "When I wanted to try a drug at a state hospital," recalls Callaway, "I went around with a cart and syringe and asked patients if they would mind if I could try something on them. There was no paperwork involved. Some doctors didn't even bother asking permission. That's just the way things were done. It was totally lax in regard to research practices."[70]

Medical research had a manifest offensive and defensive focus. John D. Marks, whose landmark book *The Search for the Manchurian Candidate: The CIA and Mind Control* exposed the CIA's longtime attraction to the black arts, including medical experimentation, argues that the agency "quickly realized that the only way to build an effective defense against mind control was to understand its offensive possibilities. The line between offense and defense—if it ever existed—soon became so blurred as to be meaningless." According to Marks, every CIA document stressed goals like "controlling an individual to the point where he will do our bidding against his will and even against such fundamental laws of nature as self-preservation."[71]

Through highly secret code-named projects, the agency explored a raft of mind-control and mind-altering potions and techniques. Hypnotism and LSD would get a thorough workout, and under the direction of the enterprising Sidney Gottlieb, the CIA's Technical Services Staff would experiment with a variety of chemical and biological agents. Various funding agencies provided cover for CIA financing that greased the skids for elite professors and renowned academic institutions to embark on some hair-raising experimental endeavors. Nearly seven dozen institutions took CIA money, many of them for the purpose of plying American citizens with hallucinogenic cocktails and more.[72]

Robert Hyde, Carl Pfeiffer, and Harold Abramson are just a few of the academic heavyweights who signed research contracts with the CIA. Some of their projects were cruel, if not criminal. Dr. Harris Isbell, for example, the director of the Addictive Research Center at Lexington Federal Prison, had a large stable of subjects to choose from and a steady supply of bufotenine, rivea seeds, scopolamine, and other concoctions to explore.[73] One of his CIA-funded experiments kept prisoners on daily doses of LSD—the crown jewel of the mind-control treasure chest—for over two and a half months. The Lexington experiments are as shocking a display of cavalier and dangerous research as one is likely to find. The fact that Isbell's test subjects were nearly all incarcerated black drug addicts sent to prison by American courts for daring to use illicit drugs adds a poignant note of irony to the situation.

Dr. Ewen Cameron, the godfather of Canadian psychiatry and the president of the American Psychiatric Association and the World Psychiatric Association, concocted some of the most chilling and dangerous psychological experiments in the name of science with the encouragement and support of the CIA. Supposedly motivated by a desire to find a cure for schizophrenia, Cameron orchestrated a hellish potpourri of depatterning techniques that left his patients—victims is a more apt description—in what he referred to as a state of "differential amnesia." One of his critics likened the process to the "creation of a vegetable."[74]

Cameron's smorgasbord approach of LSD, electroshock, sensory deprivation, and psychic driving may have left subjects bereft of their sanity and practically mindless, but both he and the CIA, which funneled him monthly checks, were pleased with the "direct, controlled changes in personality" he was able to elicit at McGill University's Allan Memorial Psychiatric Institute. As Marks concludes of the famed doctor who left no tool unused in his devilish physician's bag, "By literally wiping the minds of his subjects clean by depatterning and then trying to program in new behavior, Cameron carried the process known as 'brainwashing' to its logical extreme."[75]

Despite witnessing the devastating results of Cameron's "treatments" on his father at Allan Memorial Psychiatric Institute, Dr. Harvey Weinstein entered the psychiatric profession with a purpose: to better understand what his father had endured and discover "what kind of man would experiment on vulnerable patients?" *Father, Son, and CIA,* his graphic account of a parent's tortured existence as a human guinea pig and the physician whose "missionary zeal" knew no bounds, is a sobering look into the Cold War era's medical excesses. Weinstein's book also explores how a medical profession designed to do no harm could occasionally partake of "evil undertakings when the motivation can be reframed so that the outcome is couched under the rubric of 'for the greater good.'"[76]

Henry Murray, another iconic figure in American psychology, also fell victim to Cold War research exuberance. Viewed by many as the conduit for the acceptance of European personality and clinical theories into the American academy, Murray did biochemical research at the Rockefeller Institute before abandoning the hard sciences for psychology. Prior to World War II, he would go on to develop the Thematic Apperception Test, which was used to assess people's personalities, and became a consultant to the Office of Strategic Services (the CIA's predecessor) after the war to further perfect a methodology that would test "a recruit's ability to stand up under pressure, to be a leader, to hold liquor, to lie skillfully, and to read a person's character by nature of his clothing." His system would become a fixture in the OSS—and "the first systematic effort to evaluate an individual's personality in order to predict his future behavior."[77] Years later, during the climate of fear surrounding the Cold War, Murray would embark on related personality studies that would intentionally stress his Harvard student-volunteers. See chapter 9 for a discussion of his results.[78]

Equally unsettling is the fact that many Americans were unwittingly incorporated in postwar radioactive material studies that were designed to gauge the health effects of plutonium, a dangerous radioactive element with which hundreds of weapons scientists and Manhattan Project personnel were coming in contact. Shockingly, approximately 4,000 human radiation experiments would follow over the next few decades. Most would involve radioactive isotope investigations that tagged certain elements, such as iron, calcium, and iodine, used as measuring devices in an assortment of uptake, metabolism, and blood studies. In addition to hospital patients, test subjects included soldiers, prisoners, psychiatric patients, and average citizens. Infants and children were also often sought out by researchers as desirable test material.

The research establishment received a long-overdue wake-up call in 1966 when a Harvard anesthesiologist by the name of Henry K. Beecher published an article in the *New England Journal of Medicine* titled "Ethics and Clinical Research." The six-page broadside detailed nearly two dozen research experiments that endangered the health or life of their subjects and that had been done without informing them of the risks involved or obtaining permission for such endeavors.[79] As medical historian David J. Rothman has subsequently written, the Beecher article became "a critical element in reshaping the ideas and practices governing human experimentation."[80]

Beecher was no radical; he had no desire to cripple the medical profession or damage the reputation of his colleagues. But the real "disservice to medicine," he argued, would be to remain silent and allow the "continuation of the practices" that could prove far more injurious to the profession's reputation.[81] According to Beecher, most of those used as test subjects in the problematic protocols he cited were institutionalized and in one way or another unable to give informed consent. Newborns, mentally retarded children, charity patients, and soldiers in the military were easy prey for zealous researchers. The lives of many of the subjects were put in harm's way for the convenience of investigators.

Though his article did not identify the doctors involved, the institutions that employed them, or the funding agencies, over the years other scholars would piece that information together. Elite medical schools such as Harvard, Emory, Duke, New York University, and Georgetown made up the bulk of the twenty-two protocols he described, and funders included agencies such as the Atomic Energy Commission, the Public Health Commission, Parke-Davis, and Merck.[82] With this article, the best, the brightest, and the most influential researchers and training grounds had been put on notice.

Beecher's article didn't precipitate instantaneous reform, nor did a longer treatise by British physician Maurice Pappworth the following year, but they were clear and widely heard warning shots for an ethically lax and comfortable profession that had been imbued with a self-congratulatory, utilitarian spirit.[83] Change would come slowly, then quicken in 1972 with revelations about the Tuskegee syphilis study, but the new safeguards would be too late for the vulnerable American "volunteers" who had already been victimized.

# FIVE

# VACCINES

## *"Institutions for Hydrocephalics and Other Similar Unfortunates"*

IT WAS FEBRUARY 1973, AND THE MONTHLY MEETING OF the Pennsylvania Association for Retarded Children (PARC) at a Harrisburg hotel had just ended. Pat Clapp, the president of the statewide group, was shaking hands with board members and discussing potential issues for the next meeting when a woman approached her with a pointed question: "How could PARC have condoned using retarded children as test subjects for medical experiments?"

Clapp was aghast at the accusation. As the mother of a child with Down syndrome and a dedicated advocate for those with disabilities, she had risen to a leadership position through hard work, fighting for institutional reform and calling for change whenever and wherever possible. She was not one to blithely approve the use of disabled children as raw material for experimentation.

Clapp listened carefully and was stunned by the woman's story. Visibly upset, the woman claimed that her child had been used like a laboratory guinea pig in an experiment at Hamburg, a residential center for the retarded near Reading in Berks County.

"They're injecting live virus into children up there," she said. "My child was injected with a meningitis virus, and no one asked my permission or informed me what was going on." The woman went on to describe the situation at Hamburg and remained adamant that she had not been asked for nor had she given her permission for her child's participation in any such research.

When Clapp asked if she had any proof, the woman showed her a letter she had just received from a Dr. Weibel in Philadelphia. Incredibly, the letter stated her child had already been used in an experiment and that they were now—after the fact—asking for her permission. Equally shocking was a statement in the letter claiming that PARC had approved the investigational exercise.

"That's when the lid flew off," recalls Clapp. "I was appalled. I had never heard of medical experiments on children in these state centers, and I knew the PARC board had never discussed such an issue, much less given its blessings. I immediately started to inquire if the allegations were true and what was going on at Hamburg."[1]

Clapp had no personal knowledge of the facility, but she was not unfamiliar with large underfunded and understaffed state institutions that held those who were referred to at the time as "retarded," "feebleminded," and "defective" individuals. She would learn that Hamburg had once been the Charles H. Miner State Hospital, a sanitarium that cared for people with tuberculosis. After that disease had been conquered, the hospital was closed in 1959, only to be reopened a year later as the Hamburg State School and Hospital. More than 900 patients with myriad physical and mental maladies now resided there.

The next morning at her home in Pittsburgh, Clapp began calling everyone she knew in the mental health community, state government, advocacy arena, and the media. Now she had her own list of questions. Were the woman's claims accurate? Were children really being used as test subjects at Hamburg without researchers gaining parental permission? Were other institutions in the state system doing similar medical research without parental permission?

"I was on the phone at 8:30 a.m. and talked to Eleanor Elkin, our national president, explaining what I had just learned," recalls Clapp. "She was horrified, as was everyone else I talked to. I stayed on the phone the entire day calling people like Helene Wohlgemuth, Pennsylvania's public welfare secretary; Gunnar Dybwyd, a noted expert in retardation issues; and Henry Pierce of the

*Pittsburgh Post-Gazette.* Many people were outraged and energized by what had occurred. Parents were being taken advantage of and children were being used." Recalling what many years later grew into a multifaceted campaign to end experimentation in Pennsylvania's institutions, Clapp told us with a hint of pride, "We really went after the thing."

Within days, Pat Clapp was leading an effort that would result in numerous changes in Pennsylvania's mental health system. As she and her colleagues accumulated information, they were shocked to learn that the experiments at Hamburg were not an isolated incident. Other institutions, such as those at White Haven and Laurelton, were equally guilty of turning their charges over to researchers who were perfecting new vaccines for an array of diseases. And the parents of the children in those institutions as well as the PARC board knew nothing about it.

One of Clapp's more revealing conversations was with Dr. Robert Weibel, the University of Pennsylvania pediatrician who was behind the vaccine studies underway at Hamburg. "Weibel admitted he was doing research," recalls Clapp. "He never denied it. He said twenty children were being tested with a new vaccine. I asked him how he could do this to children—most of them between three and ten years old. But Weibel saw nothing wrong. He defended his actions. He said another doctor linked to PARC had given him permission, and he didn't feel it was necessary to notify the children's parents. But I continued to press the issue and asked how he could do this to defenseless children with an assortment of mental and physical disabilities, and he replied, 'This makes their lives worthwhile. They'll be making a contribution to society.' I was a little bit taken aback by his answer and then asked him how he could tell if the child receiving the vaccine was in pain or danger. He calmly said, 'We take their temperature.' That really burned me up," said Clapp. "That's when the fire was ignited."

A citizen-activist who had worked her way up into a statewide leadership position in the mental health arena, Pat Clapp was no novice. She understood community and issue organizing as well as the levers of power in state politics. She also understood how doctors and institutions had taken advantage of the parents of children with disabilities. "Parents had no place to turn," recalls Clapp in a recent series of interviews. "There weren't many places or programs where you could place your child. Women were afraid to say they had a child with a disability. It was an embarrassment, and most parents were incapable of taking care of a child with severe disabilities. You were usually told to put your kid in an institution. This was the best alternative, you were often told. It was the medical model at the time, and parents had little choice. They had no voice. Everyone accepted things as they were back then. Parents were desperate. They needed institutions to take care of their children. If a doctor told you something, you believed it. You did what he said. No one doubted doctors; they were authority figures. No one challenged the medical profession back then."

Times were changing rather quickly, however, and activists like Pat Clapp were in the forefront of the movement. The startling revelations of the controversial Tuskegee syphilis study, in which hundreds of black men in Alabama were studied but went untreated, were only months old, and similar but less well-known instances of medical exploitation were being unearthed in cities, towns, and hamlets across the country. Disclosures of vulnerable populations incorporated in clinical trials were making headlines on a regular basis. Clapp and her coalition colleagues helped foment Pennsylvania's contribution to this growing list of medical infamy. Public protests, letter-writing campaigns calling for a state ban on research tests, and the demand for legislative hearings were all put on the organizing agenda.

Newspapers in the Keystone State devoted considerable space to the controversy. Pennsylvanians were discovering that unethical medical experimentation wasn't just an ugly story from the Deep South; abuse was occurring in their own backyard. "PARC Panel Asks Drug Test Probe," "Experiments on Humans Denied Here," "State Kills Testing of Meningitis Shots," and "State Halts Use of Retarded as Guinea Pigs" were just a few of the newspaper headlines during the spring of 1973.[2] "Once it hit the newspapers," said Clapp, "things really blew up."

"The state will not allow Dr. Weibel, University of Pennsylvania pediatrician, to resume tests of a meningitis vaccine on a group of retarded youngsters before June 18—in effect killing the project," wrote Henry W. Pierce, one of the *Post-Gazette* reporters Clapp had contacted.[3] The article continued: "Edward Goldman, state commissioner of mental retardation, yesterday said the state would insist that Dr. Weibel adhere to a ban on human experimentation in all state institutions."

Though Weibel would vigorously appeal the decision, the authorities were not about to relent; Governor Milton Shapp's administration had been caught unaware that their facilities were being used as clinical test sites. The flood of negative newspaper stories and critical public comment was embarrassing.

As the *Post-Gazette* framed it, "Dr. Weibel's meningitis vaccine, which he was testing under contract with Merck, Sharp & Dohme laboratories, was one of several being tried on retarded children at White Haven and Hamburg State School and Hospital in Eastern Pennsylvania." Weibel admitted that others were being "tested by doctors at Rockefeller University and the National Institutes of Health" and that they had previously worked on a variety of measles vaccines. The current vaccine, said Weibel, "already had been tried on military personnel . . . and large scale trials" were now necessary "before it can be approved for general use."[4]

Encouraged to opine on the legality of such an activity, the US Justice Department admitted that no statutory law forbade doctors from conducting such experiments, with or without parental consent.[5] That assessment did not please some knowledgeable observers.

One medical expert brought in to weigh the pros and cons of the case was Cyril Wecht, the nationally recognized coroner of Allegheny County. Adamantly opposed to testing vaccines on institutionalized children unless it was designed to benefit the child, Wecht argued, "Even if parental consent is provided, it still violates the individual rights of the mentally retarded. Parents don't own their children." The bottom line for Wecht was the child's welfare: unless the experiment was "clearly to the benefit of the child, parental consent won't make it legal."[6]

Wecht's take on the case might have been informed by Justice W. B. Rutledge's comment thirty years earlier that "parents may be free to become martyrs themselves. But it does not follow they are free, in identical circumstances, to make martyrs of their children before they have reached the age of full and legal discretion when they can make that choice for themselves."[7]

Wecht had earned both law and medical degrees and understood the implications of human experimentation from both perspectives. He suggested that "the Pennsylvania Medical Society appoint a special committee along with other scientific groups to begin thinking about policies involving human experiments." According to Wecht, "lay people also should participate in the review . . . with the aim of drawing up guidelines."[8]

Though the government's investigation of PARC's charges ostensibly focused on human research, a litany of other abuses occurring at state mental facilities were exposed as well. Apparently, residents at Hamburg and other institutions were being "kept heavily drugged," making it impossible for some even to move. Some were said to act more like "robots" than humans. According to one defender of the practice, the institution was so overcrowded with "severely or profoundly retarded" individuals that the use of "psychotropic drugs" was the only way to keep the facility functioning.[9]

Equally disturbing revelations were unearthed at the Polk State School in northwestern Pennsylvania. When Helene Wohlgemuth, the state welfare secretary, made a surprise visit, she was confronted with a host of unseemly practices. One of her more unpleasant discoveries was the presence of "five-by-five wooden pens used to cage disorderly patients."[10] It was explained to her that in lieu of using straitjackets, a superintendent in the 1950s instituted the "playpens" as a control mechanism. Some were "12-by–12 feet, five times the size of those used with slats but without a top."[11] Whatever the size of the cages, however, Wohlgemuth was so disturbed by the institution's resorting to such devices—particularly when the entire system was under such public scrutiny—that on April 16, 1973, just three days after her visit, she wrote James H. McClelland, the physician in charge of Polk, a letter of dismissal. Among the reasons she listed were "cruel, degrading, and inhumane conditions" along with "severe and chronic deficiencies" concerning the training of professional staff.[12]

PARC's successful campaign to end unauthorized medical research in Pennsylvania's mental institutions was a significant victory for health advocates and critical for the establishment of principles such as parental permission for the protection of underage test subjects. But the triumphs were late—over a quarter century had passed since the Nuremburg Code was drafted. During that time and certainly for many years prior to 1947, children residing in institutions were routinely used in some of the greatest scientific quests of the twentieth century.

**TWENTY YEARS EARLIER**, two institutions for the intellectually and physically challenged in Pennsylvania—the D. T. Watson Home and the Polk State School—had played a critical role in the discovery of a long-sought vaccine to combat infantile paralysis. The Watson Home, on the outskirts of Pittsburgh, was a small upscale facility for the care of children with polio. Formerly the country estate of a successful Pittsburgh lawyer, it opened in 1920 as a residence for "crippled or deformed" children.[13] The Polk State School, situated eighty miles north of Pittsburgh, was opened in 1897 and was originally known as the State Institution for the Feeble-minded in Western Pennsylvania. Located just outside the town of Franklin, in Venango County, the facility would grow in size and provide care for both mildly and profoundly retarded children and adults.

As one of a growing list of virologists and microbe hunters seeking to conquer the dreaded disease, Dr. Jonas Salk was anxious to further explore the prevention of poliomyelitis and the creation of a viable vaccine. He had already spent considerable time experimenting with live and killed poliovirus, various adjuvants to shock the immune system, and different methods to inactivate the virus. In all of these experiments, he had used monkeys as test material. He had learned much, but the next step for Salk, according to historian David Oshinski, "was the big one: human testing."[14] But where does one go to test a new and potentially dangerous potion that could result in paralysis and death?

Salk had been introduced to the answer just a few years earlier during the war while doing influenza research in the army. The Ypsilanti State Hospital in Michigan was home to a wide assortment of mentally ill and retarded patients. Salk, along with several other researchers, took over a ninety-six-man ward, injected half the men with an experimental vaccine, and then "exposed [them] to infection by inhalation of a strain of Type B influenza virus."[15] It would certainly not be the last time Salk utilized such institutions.

Described by Oshinski as "eager, confident, aggressive"—traits not uncommon to the great microbe hunters—Salk in 1950 told one colleague in the quest for a viable polio vaccine that "I think that the time has come for these experiments to be carried out in man." Salk knew where subjects could be rounded up for risky experimental work with little fanfare: institutions for children or prison inmates. Salk, in fact, had already laid the groundwork for such a venture. "I have investigated the local possibilities for such an experiment and find . . . there are institutions for hydrocephalics and other similar unfortunates. I think we may be able to obtain permission for a study."[16]

Salk had specific institutions in mind and knew which superintendents would be amenable to his using "inmates" as test subjects.[17] One of them was Dr. Gale H. Walker, the head man at Polk. Salk knew how to appeal to men like Walker and sometimes used intermediaries to bolster his case. It was only slightly surprising then that Superintendent Walker would become an avid supporter of the Pittsburgh virologist's request. As he began his letter to Pennsylvania's secretary of welfare, Walker frankly stated, "During the past six months I was approached by several prominent practicing physicians who were anxious for me . . . to meet with Dr. Jonas E. Salk, Research Professor of Bacteriology, School of Medicine, University of Pittsburgh, and the Director of Bacteriology, School of Medicine, University of Pittsburgh, and the Director of the Virus Research Laboratory there." Walker was requesting permission to enter into a "collaboration" with the University of Pittsburgh School of Medicine and the National Foundation for Infantile Paralysis in a proposed field study at the Polk State School dealing with "the administration of poliomyelitis vaccines to patients in our institutions."[18]

Sounding more like a personal agent than the head of an understaffed and overcrowded state institution, Walker would go on to illuminate Polk's attributes as a test site: its stable, long-term population, controlled diet, and "relative isolation from the general community." He admitted his "intensely favorable" reaction to the request and his firm belief that "in no manner could the charge of using institutional patients for guinea pigs be leveled at us." Walker also informed his boss that he had "already taken the liberty of approaching several patients' parents" with the idea and received positive feedback.

A few weeks later, the Pennsylvania commissioner of mental health gave Walker his answer. He admitted that the commonwealth had a "vast resource of clinical material on hand" for these types of "research projects" and that he had "frequently" acceded to the research requests of "the superintendent of Pennhurst" and "University of Pennsylvania pediatricians." But the commissioner was clear that no such endeavor should place patients at risk. None could be "considered an assault upon a patient" nor could any patient be placed "under jeopardy of health."[19] In the end, Salk would gain access to the wards of an overburdened state facility and initiate his polio virus experiment. Risk was certainly involved, but it would not be the cherished and pampered students of elite Pittsburgh area prep schools like Sewickley Academy and Shady Side Academy who would shoulder the dangerous burden of testing a new vaccine; it would be those who were still occasionally referred to as "feebleminded," "morons," and "mental defectives."

The leap from animal experimentation to clinical trials with humans was a chasm that only the most driven and single-minded investigators hurdled without trepidation. The jump was made that much more palatable, however, by knowing that near-humans—the insane, "mental defectives," criminals behind bars, and other lesser members of society—would be used as test material. Still, some were nervous about exposing even institutional flotsam and jetsam to potentially paralytic and deadly diseases. Polio wasn't the flu; it could easily destroy a person's life.

Isabel Morgan, for example, a talented Johns Hopkins researcher and the daughter of accomplished biologists, one a Nobel Prize winner, was further along than Salk in developing a killed poliovirus in the late 1940s, but she jettisoned the effort when she married and became a housewife and stepmother. Close friends knew, however, that not only marriage had impeded Morgan's progress in the laboratory. She had often expressed a decided reluctance to use humans as lab rats. She was concerned that she might end up paralyzing test subjects rather than discovering a viable vaccine.

Others, however, were willing to expose humans to a potentially harmful germ, dangerous substance, or painful procedure. Hilary Koprowski was such an individual. The young Polish scientist had fled Europe after the Nazis invaded his homeland in 1939 and first settled in Brazil with his wife, Irena, also a scientist. Hilary procured a job working for the Rockefeller Foundation doing research on yellow fever. By the end of the war, he and his family had migrated to America, where he accepted a position with Lederle Laboratories as a researcher at their Pearl River campus in New York. One of his first assignments was the development of a live-virus polio vaccine.

Possessed of old-world charm, a connoisseur of gourmet cooking, and an accomplished musician, Koprowski was also bright, creative, and bold. He was a definite risk taker; no one had ever accused him of lacking self-assurance. His research game plan was quite simple: "I decided to first attenuate [weaken] poliomyelitis virus, then to find out whether it was possible to develop a vaccine that would replicate in human gut without causing signs of disease, and then to immunize people by feeding them this vaccine." Koprowski planned on his creation being the world's "first oral vaccine."[20]

After spending several years attenuating the virus in mice and rats, he decided to test it on rhesus monkeys. Koprowski was elated: "Not a single monkey became paralyzed." Better yet, they "developed antibodies and were resistant to challenge with a virulent strain of the virus." Further experiments on chimpanzees followed, with Koprowski believing he had discovered a viable polio vaccine. Following in the tradition of some great medical researchers who abided by an unwritten code that they be the first to test potentially dangerous elixirs, Koprowski and his loyal assistant, Thomas Norton, imbibed a "polio cocktail" consisting of ground-up rat spinal cord and mashed brain tissue. Both survived the unappetizing beverage.

Two years later, Koprowski confronted what he referred to as their next problem: "whom we would vaccinate." According to our interview with Dr. Koprowski in 2009, the answer came in the form of a timely request from Dr. George A. Jervis, a physician at Letchworth Village, a state institution for "abnormal children" near Lederle's Pearl River campus. These were "not experiments" in the traditional sense, claimed Koprowski, but a response to "a plea by authorities of the institution. Children were eating feces and throwing feces all over the place and at each other. There was contamination throughout the entire facility, and they believed the children were at risk of contracting polio." Dr. Jervis, whom Koprowski called "'a close friend,' pleaded with us, 'Please try the vaccine at Letchworth.'"[21]

In his account of that first human trial, Koprowski realized he would "never get official permission from the State of New York." He decided to bypass state authorities and just "ask permission from the parents of these children. On February 27, 1950, the first human subject was immunized with poliomyelitis virus by drinking an emulsion of cotton rat brain and cord."[22] Oshinski's account of the historic incident leaves open the question whether "Jervis got consent from the children's parents or simply took on the responsibility himself." That first subject, a young "boy [with] no antibodies," was chosen, and after several weeks, Koprowski, Norton, and Jervis "increased the number of children to 10, and ultimately ended with 20 children."

That promising first step wasn't revealed until March 1951, when Koprowski attended a conclave of the National Foundation for Infantile Paralysis in Hershey, Pennsylvania. A collection of highly accomplished academic and university hospital researchers were in attendance, including Joseph Stokes Jr., David Bodian, Thomas Francis, Jonas Salk, and Albert Sabin. The little-known Lederle researcher was definitely considered a neophyte in such elite company, but his presentation quickly captured the attention of the august assemblage. Being the first to use a live poliovirus on children qualified as something definitely noteworthy.

Most of the attendees were stunned, with Sabin practically apoplectic. According to Koprowski's account, "Sabin was quite vociferous at the meeting.

Sabin questioned my daring. How did I dare to feed children live poliovirus? I replied that somebody had to take this step. Well, he turned round and round saying, 'How do you dare to use live virus on children? You are not sure about this, you are not sure about that, you may have caused an epidemic.' Even kind Dr. Stokes . . . asked whether I had checked the possibility that the Society for Prevention of Cruelty to Children would sue me for what I had done."[23]

Over six decades later, Koprowski chuckles at "the great polio war" and the intense competitiveness of his peers in the virology community as to who was going to be the first to claim the title as the microbe hunter who bagged polio. He ruminated over every scientist's dilemma: "doing something when you don't know the outcome." He said that "there was a fight over everything" and believes Sabin's attack at Hershey had more to do with "jealousy" than with any ethical rules Koprowski may have breached. Sabin "was not a nice character," says Koprowski. "He was a fine scientist, much better than Salk, but human nature is not nice. We are jealous of each other and that motivates us to do things. It was an unethical family." He wonders why they paid little to no attention to "Dr. Stokes vaccinating female prisoners" in a New Jersey penal facility and using their "newborn children as research subjects."[24]

As for ethical restrictions during the early postwar years, Koprowski says there were few and that he was not aware of the Nuremberg Code. While others at the time may have been equally uninformed, they still had some personal guidelines concerning who could be used as test subjects.

Thomas Rivers, a major figure in twentieth-century virology and a senior scientist at the Rockefeller Institute, was a shocked participant at the Hershey meeting. A decade and a half later, he offered his assessment of the Koprowski/Letchworth Village experiment. He was clearly not a fan of the commercial scientist or his methods. "First," recalled Rivers, "I didn't think that the safety tests that Dr. Koprowski had done were anything to write home about, and second, I personally did not approve of using mentally defective children for such a test with inactivated vaccines, and what Koprowski wanted to do was not unusual—you might even say that it was standard practice."[25]

Rivers held strong views on the subject of using institutionalized children as research material and was out of step on the issue with many of his colleagues at the time. Of the early Cold War years, Rivers said, "Some people at the Public Health Research Institute of New York wanted to test what was then a new typhus vaccine on some mentally defective children in Letchworth

Village, and they found me bitterly opposed. . . . I think that if someone wants to use adults as volunteers to try out a new drug or vaccine, that is perfectly all right, provided that the adult had been told about the nature of the disease he is exposing himself to, has been completely informed about the nature of the agent he is to receive, and been told the chances for success or failure."[26]

Further commenting on the subject of human volunteers, Rivers believed there were two classes of research volunteers: either prisoners in state or federal institutions or scientists themselves. He admitted,

> I don't even know that you can actually call a prisoner a volunteer. I believe that although prisoners are usually told that they will get nothing out of volunteering as guinea pigs, deep down they believe that they may get a commutation or reduction of their sentence. That's perfectly all right: the point is prisoners are generally adults who can weigh the pros and cons of submitting to a test, and if they arrive at a decision to participate in a test, it's a decision or judgment that they have made. It's not made for them. An adult can do what he wants, but the same does not hold for a mentally defective child. Many of these children did not have any mommas and papas, or if they did their mommas and papas didn't give a damn about them.[27]

Nevertheless, Koprowski's Letchworth Village experiments and others with larger numbers of children at Sonoma State Hospital, an institution for the mentally ill and mentally challenged in California, ultimately were viewed as a turning point and a courageous step in the difficult campaign to conquer the dreaded paralytic disease. Time and events greatly impacted scientific research in these early days of the Cold War; there was a sense of urgency, and medical breakthroughs were all important. Salk himself would soon be doing experiments at Watson and Polk—research that didn't attract the negative criticism Koprowski's had received. In 1952, for example, the British journal *Lancet* chided Koprowski for suggesting that children in his studies were actually "volunteers."[28] Salk managed to avoid such sniping commentary. Even Sabin, one of Koprowski's most ardent detractors that day at Hershey—and someone who, like many others, had been influenced by Paul de Kruif's *Microbe Hunters* in their youth—would go on to push the ethical envelope.

Instances of cavalier research would become increasingly common. In December 1951, at a meeting of polio researchers in New York City, for example, Dr. Howard A. Howe, a distinguished virologist, presented a paper on the need for additional immunology studies and suggested a "pilot experiment" that would inject fifty to a hundred children—"presumably mentally deficient"—with a trivalent (three strains of polio) vaccine. If that exercise proved successful, the test population would be enlarged "to include 1,000 normal children in the age range of 1–3 years."[29] A year later, he would write an article discussing antibody response of chimpanzees and humans to a polio vaccine. The humans for his study? "Eleven bedridden children in the second to fifth years of life" from the Rosewood Training School in Owings Mills, Maryland. All, according to Howe, "were low-grade idiots or imbeciles with congenital hydrocephalus, microcephaly, or cerebral palsy." Howe went on to assure journal readers that "despite these handicaps their physical condition was . . . good," thereby making them fine test subjects.[30] There is no record of opposition at either the New York gathering or from journal editors regarding Dr. Howe's belief that "defective children" were the chimpanzee's closest relative, at least from a clinical trial perspective.

As Rivers candidly observed, use of "defective children" and other disadvantaged populations had become "standard practice" among medical researchers. For decades, similar refrains had been heard at scientific meetings. At one 1936 high-level meeting of the President's Birthday Ball Commission on Infantile Paralysis in Baltimore attended by both Rivers and Paul de Kruif, who was the group's secretary, a dozen influential doctors discussed the belief that only "through studies on children" would a victory finally be achieved.[31]

Physicians in the 1930s were working under the mistaken belief that polio was an airborne disease and that the virus could enter through the nasopharynx.[32] An assortment of nasal sprays and nasal injections were being explored as possible preventives. When the discussion at the Baltimore meeting gravitated to the "difficulties of immunization" and potential test populations for trial runs, de Kruif offered a suggestion: "Could there not be a test set up where half of some age group would be given picric acid [a potential preventive], and then control some colored material?" Suddenly, another doctor, offended by the comment, interjected, "I do not think we should discuss it here."[33]

Though the practice of using vulnerable populations was commonplace—the Tuskegee syphilis study was in its infancy at the time—boldly suggesting such an idea at a public meeting was not encouraged. De Kruif's comment was not the only problematic suggestion uttered at the high-level medical conclave. One doctor ruminating over the need for a poliomyelitis trial openly wondered about the possibility of using two different test groups and "two different solutions and us[ing] them alternately," an experiment he even thought might lay bare "questionable ethics."

The Baltimore brainstorming session demonstrated other examples of utilitarian bias and blatant self-interest regarding large-scale experimental vaccine trials. The most egregious was probably the lengthy discussion regarding sources for test subjects; one creative suggestion was "to get up an experiment" that would target the nation's many "Infant Welfare Clinics." As one participant forcefully stated, "We have 2,000 children in our own welfare clinic between two weeks and six years of age. Throughout the country there are hundreds of thousands of children in this same relation."[34] The logic was incontrovertible. Researchers needed children as test subjects, and welfare clinics had them in abundance.

When Rivers offered a cautionary note about such an initiative and expressed a fear of the public's learning about a dangerous experiment involving their children, he was met with the reply: "The group of people with whom we deal in the Infant Welfare Clinic would not be affected by this advance publicity, and of course I would not encourage any of them. We will tell them, now is the time to take your baby for a tuberculosis test, diphtheria test, etc."[35] Obviously, deception and mendacity toward uneducated and economically disadvantaged people was not a problem for some physicians.

As Dr. Donald Armstrong said in support of the economic advantages of using Welfare Clinic children, "It is a group easy to handle, already corralled, who could be treated with 1/12 of the expense of the population that you could go out and set up for yourself."[36]

The brazenness of Depression-era polio fighters is even more fascinating considering that one of the worst scientific fiascos of the decade was still in the news. In the race to be the first to conquer the dread disease, several doctors had begun vaccinating children with their new anti-polio concoctions only to discover what had always frightened Dr. Isabel Morgan; instead of protecting humans from disease, they were giving it to them. Paralysis and death were the result.

In 1935 Dr. John Kolmer, a pathologist at Temple University in Philadelphia, William H. Park, a highly respected bacteriologist at New York University, and his protégé, Maurice Brodie, claimed that they had discovered safe and effective vaccines for infantile paralysis. Brodie, determined to make a name for himself by discovering an effective polio vaccine, had convinced his esteemed mentor that he had done so through a "virus containing emulsions of infected monkey spinal cords and inactivated with formaldehyde." After successful trials with his killed virus on monkeys, Brodie, Park, and a few lab assistants inoculated themselves with the vaccine and survived with no ill effects. Dispensing with further trials and not shy about his presumed success, Brodie went to the press and extolled the significance of his accomplishment. Newspaper headlines sang the praises of the doctors for providing hope to millions of parents who feared the ominous shadow that annually crossed the American landscape leaving thousands of paralyzed and dead children in its wake. The scourge had seemingly been conquered.

Kolmer, however, was also in a boastful mood. His vaccine, different from the Brodie/Park version, was a "live but devitalized" virus that had been further attenuated through further laboratory treatment.[37] Moreover, it had been tested on twenty-two children in addition to himself and his own two children. Newspapers were quick to announce the breakthrough: "New Infantile Paralysis Vaccine Is Declared to Immunize Children" and "Will Give Children Paralysis Vaccine."[38] The press stirred up the competitive rivalry to the point where newspapers carried progress reports resulting in each team's attempting to outdo the other.[39]

Not everyone was caught up in the hoopla, however. Thomas Rivers remained unconvinced. By October 1935, Brodie's vaccine had been given to about 8,000 people and Kolmer's to another 12,000, but signs of concern had surfaced. Some children had contracted polio where no outbreak had occurred but vaccine trials had. Rivers believed that the vaccines had never been proved safe and that some people—at least eight by his count—might have contracted polio from the inoculations. As he told the press, "Information in my hands about the time and circumstances of these eight cases make it imperative that Dr. Kolmer show his vaccine is absolutely safe."[40]

Other eminent medical scholars would weigh in on the controversy. The legendary Dr. Simon Flexner, the director of the Rockefeller Foundation, argued, "No adequate evidence exists to support the claims" that the Kolmer and Park/Brodie vaccines "were effective agents against infantile paralysis."[41] Flexner strongly criticized both methods, which he considered unsafe, and urged caution.

Park and Brodie were not intimidated by Flexner's warnings. "We are giving definite immunity with our vaccine treatment," argued Brodie. "The fact that our . . . virus of infantile paralysis is killed by formalin obviates any possible danger of giving the disease to any child who might not be susceptible to it. There is absolutely nothing to lose and everything to gain by continuing our tests."[42]

Evidence was quickly proving contrary to what the doctors were claiming. Instead of protecting people, the vaccine experiments themselves were giving people polio. There was confirmed data showing "at least 12 vaccine-associated cases and six deaths."[43] By the end of 1935 and after much professional embarrassment, the race for a new polio vaccine was seen as a "foolhardy venture" that destroyed lives. Medical researchers in general, and certain doctors in particular, were criticized for "lack of restraint, poor judgment" and the use of innocent citizens as laboratory test subjects. The research debacle left the once-esteemed William Park with a damaged reputation and Brodie and Kolmer considered persona non grata among their peers. The research community itself took a serious hit; ego and the pressure to be the first with a polio vaccine had led to mass vaccinations before their safety was ensured. Worse yet, infantile paralysis was still out there.

During the early 1950s, however, recollections of that terrible medical faux pas had begun to fade, but the siren call of personal success and scientific triumph was as strong as ever. A new generation of young and aggressive microbe hunters had come on the scene, and they knew where experimental vaccine trials could be tested cheaply and quietly.

**ONLY A FEW YEARS AFTER** Koprowski and Salk began their studies at Letchworth Village and Polk State School, Dr. Saul Krugman was invited into the Willowbrook State School on New York's Staten Island. The large, perpetually overcrowded institution, home to over 4,000 children with a variety of mental and physical disabilities, was suffering from an outbreak of infectious hepatitis. Krugman, a pediatrician at New York University Medical School, could not help but connect the scores of children acquiring the liver disease with the facility's horrid sanitary conditions and the children's propensity to languish in their own filth.

Krugman would perform experiments at the institution for well over a decade. His research would make a significant contribution to science and increase our understanding of infectious disease, but his studies would also become some of the most controversial human experiments of the last half-century. For decades, defenders and detractors of Krugman's research would square off at forums, in journals, and on op-ed pages to illuminate what was right or wrong with feeding virus-laden stool to mentally retarded, indigent children on the wards of a state mental hospital. The ethical issues at stake were so complex and convoluted that over time some ardent opponents found themselves becoming enthusiastic champions of Krugman's work.

Dr. Henry Beecher's controversial 1966 *New England Journal of Medicine* article drew uncomfortable attention to twenty-two dubious clinical studies that were said to put patients and test subjects at risk.[44] Though Krugman was not named, his research was example 16 in the article. The issue of whether it was morally right to purposefully infect a human being with hepatitis virus was heightened by the fact that the humans in question were children—mentally and physically challenged children, those most in need of care and protection in the New York State child welfare system. And their numbers ran into the hundreds.

A surprising number in the medical community, however, felt that such research could be justified. Dr. Franz Ingelfinger, for example, the esteemed editor of the *New England Journal of Medicine*, argued in 1971: "How much better to have a patient with hepatitis, accidently or deliberately acquired, under the guidance of a Krugman than under the care of a zealot who would exercise intuitive management, blind to the fact that his one-track efforts to protect the rights of the individual are in fact depriving that individual of his right to good medical care."[45]

Such arguments proved unconvincing to Krugman's detractors, who were less concerned with his credentials as a serious-minded investigator or the results of his research than with the "ethics of his studies." As Dr. Stephen Goldby wrote in the *Lancet*, "the whole of Krugman's study, is quite unjustifiable, whatever the aims, and however academically or therapeutically important are the results." Goldby would go on to voice his amazement that such ethically flawed work would be "actively supported editorially by the *Journal of the American Medical Association* and by Ingelfinger." In Goldby's opinion, it was "indefensible to give potentially dangerous infected material to children, particularly those who were mentally retarded, with or without parental consent, when no benefit to the child could conceivably result."[46]

That was the heart of the issue for Goldby and many others: "Is it right to perform an experiment on a normal or mentally retarded child when no benefit can result to that individual?" Their answer was no. As the Willowbrook study opponents believed, "it was the duty of a pediatrician in a situation such as exists at Willowbrook State School to attempt to improve that situation, not to turn it to his advantage for experimental purposes, however lofty the aims."[47]

That was the point that Beecher tried to make when he first brought up the issue of increasingly problematic research appearing in American medical journals. He was not out to lambaste or embarrass particular doctors, but the ever more frequent incidences of unethical research were becoming increasingly troubling. Beecher believed that it was critical for the profession to understand the importance of ethics. He wanted all to know that "a study is ethical or not at its inception. It does not become ethical merely because it turned up valuable data."[48]

Krugman himself responded to the attacks at various forums and in writing, usually saying that the ethics of a situation could not be divorced from the time and place in which these studies were conducted. Willowbrook in the mid-1950s, he argued, was a very special and troubled place. As a result of the unsanitary and overcrowded conditions, newly admitted children were likely to become infected with infectious hepatitis and jaundice within six to twelve months. The answer to this growing problem, said Krugman, was the acquisition of knowledge that would lead to an effective immunizing agent—a vaccine that would protect children the way the Salk and Sabin vaccines were protecting children from infantile paralysis.

As Krugman would subsequently write of that initial decision to tackle the Willowbrook hepatitis problem, "It was essential to acquire new knowledge about the natural history of this disease—knowledge that might lead to its ultimate control." After three decades of both praise and scorn for his work at the institution, Krugman remained convinced his studies were both "ethical and justifiable."[49]

In their initial "Plan of Study" for tackling the Willowbrook hepatitis conundrum, Krugman and his NYU medical colleagues sought answers to two questions: "Would gamma globulin stop hepatitis . . . in the same manner as reported by Stokes and his associates working in a similar institution?" and "Could passive-active immunity be experimentally attempted in human subjects who would later be tested for protection?"[50] The results of their early studies on Willowbrook children showed no protective effect from gamma globulin. To answer the second question, the researchers simultaneously injected children with gamma globulin and fed them virus-containing material made from the stool of six patients with hepatitis and jaundice that had been so refined that Krugman's team deemed the end product "safe to feed."[51]

According to the NYU team, they obtained permission and "gingerly" fed children three to ten years old the infected suspension in chocolate milk. The subjects were "kept in separate isolation wards of the same building," and one of the children "came down with typical hepatitis with jaundice on the 30th day." A second trial resulted in five of eleven children developing "typical jaundice with confirmatory liver tests." Desirous of "produc[ing] jaundice in a larger proportion of the recipients, 13 children were fed" an increased dosage of suspension. "This resulted in hepatitis with jaundice in 12 of 13" children.[52]

There is little doubt that Krugman and his research team produced some valuable information about the disease, but the experimental results probably provided little solace for the children now suffering from "fever, vomiting, diarrhea, enlargement of the liver, and equivocal liver function tests."[53]

The hepatitis experiments would go on into the 1970s while helping to establish two clear types of the disease—hepatitis A and hepatitis B—and pave the way for successful active and passive immunization. But those triumphs did little to negate the reaction of some to feeding infected serum to retarded children. Even as far back as the 1950s, when a lone dissenting voice commented on research practices at Willowbrook, Krugman was quick to defend his program, saying, "We, too, are keenly aware of our moral responsibilities. This study was not undertaken lightly."[54] In fact, a formidable list of sponsors and supporters including the Armed Forces Epidemiological Board, NYU's School of Medicine, the New York State Department of Mental Hygiene, and the *Journal of the American Medical Association* had bought into it. All knew about the research program; not one had a problem with it.

New allegations arose in 1964, claiming that the parents of Willowbrook students had been coerced into signing their sons and daughters over to the researchers. Due to intense overcrowding at the institution, new admissions had been curtailed—unless, of course, parents were willing to have their child become a test subject in Krugman's Willowbrook research unit, where spots were still available.

Lost in the long-running debate over the Willowbrook hepatitis experiments are some other interesting aspects that we believe deserve comment. Researchers had a rare opportunity at those institutions housing vulnerable populations that opened their doors to them: The wealth of test material combined with the presence of other health maladies, and the freedom and convenience to explore potential remedies, was an investigatory opportunity that allowed scientists to experiment to their heart's content. For example, in early 1957, not long after the NYU team started work at Willowbrook, Dr. Robert Ward, Krugman's associate, wrote the school director about a continuation of the hepatitis research that would explore everything from additional stool studies and the relationship of the newly synthesized Thorazine to infective hepatitis; polio's relationship to various "neurological disorders"; a measles vaccine trial; a study of "amino acids in the blood and spinal fluid"; and research on "new types of inborn errors of metabolism which might be correctible by diet."[55]

Obviously, Ward believed, such an expanded research program required "separate isolation units," additional "laboratory space," and more staff "to care for the patients and conduct the studies." He suggested "the second floor of Building 2" as the ideal location for the new, enlarged program. In closing, Dr. Ward thanked the director for his cooperation and wrote, "Medicine have [*sic*] profited enormously by their experiences with rubella, pertussis, tuberculosis, and hepatitis at Willowbrook." It was "in the spirit of continuing this happy relationship" that the researchers now asked for more space, more freedom to explore other scientific pursuits, and more test subjects.[56]

The state school for the mentally retarded had been transformed into an attractive investigative gold mine for NYU researchers. Extant documents from the 1950s and 1960s disclose Dr. Krugman or one of his associates writing to Dr. Harold Berman, Willowbrook's director, for permission to conduct additional research there. In the spring of 1959, for example, Krugman told Berman about Dr. John Enders of Harvard Medical School and his research into live attenuated measles virus. Enders had administered the vaccine to children in a Massachusetts institution, and Krugman thought that "these studies could be extended at the Willowbrook State School."[57]

Krugman also assisted the US military when possible. In 1960, for instance, one army medical investigator wrote Krugman for "any prebleed sera" from the Willowbrook "volunteers" for the military's own experiments.[58]

Krugman claimed a "limited supply" of his own since "our subjects are children" and therefore it is "difficult to obtain large quantities of blood."[59] We do not know whether such requests pressed the NYU researchers to increase the number of test subjects or experiments.

By the mid-1960s, Krugman was a major name in research circles with a number of notable firsts to his credit, including a pivotal role in the development of vaccines to prevent mumps and rubella and of the trivalent mumps-measles-rubella vaccine. He would send the Willowbrook directors thank you notes for their "cooperation" in permitting the "many outstanding Willowbrook studies" and assuring them that "the name Willowbrook [had achieved] a permanent place in the world's medical literature."

Such letters often were followed by others requesting permission for another clinical trial. For example, in the summer of 1965, Krugman asked permission to "conduct a Diphtheria Immunity Study" that would require at least "360 individuals, and preferably 480 individuals."[60] Less than twelve months later, Krugman was back asking permission to do another vaccine study, as the Merck Institute for Therapeutic Research was looking to build on "preliminary clinical studies . . . carried out in the Philadelphia area" into the measles vaccine.[61]

Documents disclose that the medical research program often caused administrative problems for the Staten Island facility. In fact, on one occasion in 1967, Dr. Jack Hammond, then the Willowbrook director, asked Krugman to inform an associate of the increasing burden that new research projects imposed on his custodial operations. "The additional administrative and paper work," said Hammond, "that such a [proposed] project will place on my already over-worked staff, in view of the need to obtain informed parental consent for the administration of the para-thyroid extract on an experimental basis [*sic*]." Hammond wanted the doctors to appreciate his organizational bind and "the great problem of insufficient staff to insure urine collections when I have perhaps two attendants at night for wards holding anywhere from 50 to 80 patients." Hammond was justly sensitive to "complaints of over-work" by ward and stenographic staff due to "too many research projects conducted at the request of other agencies." In short, Hammond supported Krugman's research program and was willing to endure the "sensational accusation[s] concerning research and experimentation" with these mentally challenged children that he was periodically confronted with, but the "additional administrative" burdens were a recurring headache.[62]

For Krugman, however, Willowbrook was a career maker: a huge, unfettered laboratory that allowed him to explore the hidden mysteries of viral disease to his heart's content. With a neglected and devalued population of over 6,000 residents living in close quarters and devoid of any say in the matter, Krugman and his fellow investigators had a huge repository of cheap and available human test subjects.

Great things would come of Krugman's human experiments at Willowbrook. New knowledge, groundbreaking vaccines, a river of honors and awards, and front-page *New York Times* headlines solidified Krugman's reputation as a microbe hunter in the best tradition of his forebears.[63] Among Krugman's discoveries was the fact that there actually were two strains of hepatitis, hepatitis A and hepatitis B, and his research would discover a vaccine to prevent the latter. There can be little doubt that children in succeeding years at Willowbrook and around the country were better off for Krugman's many discoveries. But the thousands of children from the 1950s to the 1970s who were "volunteered" into hepatitis, measles, mumps, rubella, and other experiments would reap no awards or headlines. Nor would Paul de Kruif chronicle their story.

**THE WILLOWBROOK CONTROVERSY STIMULATED** considerable debate and consternation over the years. In 1972, one defender argued that Krugman had "produced more information about hepatitis . . . than anyone else in the world."[64] That defender was Dr. Stanley A. Plotkin, associate professor of pediatrics at the University of Pennsylvania.

Plotkin was no research novice or blindly loyal Krugman supporter; an established scientist in his own right, he would spend thirty-one years as a member of the research faculty at Wistar Institute. He was familiar not only with doing serious clinical research but doing it in state institutions as Krugman did. Plotkin had invested considerable time in studying rubella or German measles, a disease that was generally mild in children but far more dangerous in pregnant women. A child born of a rubella-infected mother could be beset with a range of serious incurable maladies. Plotkin's discoveries would go a long way in helping to eradicate rubella in the United States.

In 1964 and 1965, America, like much of Europe, would become part of a rubella pandemic. It was estimated that there were over 12 million cases in the United States, and that thousands of infected pregnant women underwent therapeutic abortions. Plotkin would spend years researching the disease.

One of his more provocative pieces of research grew out of one individual's personal tragedy and concerned a series of rubella vaccine trials at an orphanage and institution for the mentally retarded in the mid-1960s. A pregnant woman became infected with the rubella virus and was encouraged to have an abortion. The fetus was surgically aborted and dissected, according to a subsequent Plotkin publication, and "explants from several organs were cultured and successful cell growth was achieved from lung, skin, and kidneys. All cell strains were found to be carrying rubella virus."[65] After being put through a refinement process and several passages (a process to attenuate or weaken the strain), the potential vaccine was tried on "adult mice" and "African green monkeys." Plotkin and his lab partners now believed the vaccine was ready for human trials.

According to one of his rubella publications, Plotkin chose to do his first human trials at the Hamburg State School and Hospital in central Pennsylvania in addition to "volunteer families living in Philadelphia." He would also claim to have attained full written informed consent of parents, guardians, and relevant authorities. Children in the study were said to be "moderate to severely retarded" and ranging from "4 to 13 years of age." Subsequent rubella studies revealed that Plotkin also gained access to St. Vincent's Home for Orphans in Philadelphia, where his test subjects ranged from "1 1/2 to 3 years in age." The children were either given injections of the new vaccine or kept in a confined space with those who had received the rubella vaccine where "contact between subjects was promoted."[66]

Both institutions had a history of cooperation with researchers and hosting clinical trials. The choice of a Catholic orphanage is interesting, however, considering the fact that the rubella vaccine was derived from an aborted fetus, something church officials would presumably have frowned upon. How much of that information "Sister Agape," the Mother Superior of St. Vincent's, was privy to is open to question.[67]

IN NOVEMBER 1950, Dr. Richard Capps of the Liver Research Laboratory at St. Luke's Hospital in Chicago wrote to Dr. Joseph Stokes Jr. a letter laying out an experimental venture and concluding with the optimistic line, "It seems to me that this is a very neat little project, and if it works, we'll really have something."[68]

The project Capps referred to was an experiment testing the ability of gamma globulin to prevent the onset of hepatitis. The initiative grew out of a hepatitis epidemic at Chicago's St. Vincent's Orphanage and Hospital during the 1940s. St. Vincent's was a five-story structure that housed about 200 infants and children. There was also a small ward for expectant mothers who after delivery often worked in the orphanage. Oddly enough, new student nurses were particularly hard hit with hepatitis, and Stokes was sought out for his knowledge of the disease and his assistance in eradicating it. Stokes recruited a team of Chicago liver specialists, who began mapping a strategy to test their hypothesis.

They decided to set up a special ward of twenty-five beds that would include twenty normal children, a third of whom received gamma globulin. They would then "introduce . . . two of our old chronics who still seem to be active" and then "fresh cases" some time later. An added feature of the experiment would be by "special arrangement with the Sisters . . . poor nursing technique in this one ward which, of course, is absolutely essential." An added component of the study would be "repeated skin tests" of the experimental participants as well as of twenty-five additional nurses and children in the building.

Capps, the primary designer of the project, was buoyed by the fact that "I have already talked to the Sisters, and the . . . plan is definitely feasible." He was excited by the prospect of testing gamma globulin's power to thwart infection and further determine whether "skin test material produces active immunity."

Stokes was "in accord" with the design, and Capps was ready to begin exposing children to hepatitis. But Stokes first offered up a few cautionary notes. He was concerned about an excessive number of skin tests and commented that "the new test material will not be ready until January [1951] because we want to be sure the new test material will not produce hepatitis." He underscored the point: "I think you can understand why we must be careful."[69] He did not want to be the reason for a hepatitis epidemic at the institution.[70]

**EARLY TWENTIETH-CENTURY DISEASE FIGHTERS** used an array of diagnostic techniques and methodologies to assess the health of their patients and determine what specific malady they suffered from. Tuberculosis had been a particularly incapacitating and deadly disease well into the middle decades of the last century. For many years it was detected through the use of tuberculin, a glycerinated extract of the tubercle bacillus that some believed might be a cure for the disease. Three different methods, each named after the physician who invented it, incorporated tuberculin as the analytical divining rod that would detect tuberculosis. The Calmette test dropped a solution filled with tuberculin into the patient's eyes; the Moro test injected the tuberculin into the patient's skin; and the von Pirquet test injected the tuberculin solution directly into the patient's muscle.

It was not long before the Calmette test faced growing concern, if not outright opposition. As one medical observer wrote, "The alarm had been sounded by several oculists of note" resulting in "a growing distrust and fear among ophthalmologists" about the "use of the conjunctival test." Increasingly, doctors refused to use the Calmette test: "They considered it dangerous."[71]

As the medical community debated the strengths and weaknesses of Calmette versus the other two tuberculosis tests during the first decade of the twentieth century, several researchers at the William Pepper Clinical Laboratory at the University of Pennsylvania decided to examine the merits of the three diagnostic tools. Their experimental test site: the "enormously crowded" Catholic St. Vincent's Home for Orphans on the banks of the Delaware River in Philadelphia. The experiment incorporated children under eight; most were just three or four years of age. The doctors coordinating the study were delighted that they had "absolute control throughout the entire period that the tests were being made."[72] As medical historian Susan Lederer describes the event, the Penn physicians applied the tests to over 130 young orphans, resulting in great discomfort and "serious inflammations of the eye."

Antivivisectionists were some of the first to take note of the growing practice of using children in research, and they employed the accounts of children being subjected to experimentation in their efforts to harness the aggressive practices of researchers. Their leaflets and newsletters dramatically described the suffering endured by orphans at the hands of heartless tuberculin-dropping physicians. "The little children would lie in their beds moaning all night from the pain in their eyes. They kept their little hands pressed over their eyes, unable to sleep from the sensations they had to undergo," according to one activist's account.[73] Photos of sweet little girls no more than five years old in identical dresses with colorful bows in their hair only added to the powerful message that university doctors were routinely exploiting society's least advantaged members.

Despite the criticism, other doctors also wanted to add their thoughts to the Calmette, Moro, and von Pirquet debate. Louis M. Warfield, a St. Louis physician, for example, published an article after using the von Pirquet test on over one hundred children, fifty-five of them infants. He found another fifty-one subjects between the ages of five and fourteen in an "orphans' home."[74]

Warfield's numbers were downright puny compared to Dr. L. Emmet Holt's sample of "one thousand tuberculin tests in young children" at New York City's Babies Hospital in 1909. A noted professor of pediatrics at Columbia University's College of Physicians and Surgeons, Holt was well aware of "the possible dangers connected with the eye test" and potential "scarification" from skin tests, but he moved ahead undaunted. In one journal article, he proudly claimed that the reports that "young infants do not respond to the eye test . . . was not our experience."[75]

Holt admitted "considerable experimenting" before he settled on a proper dose for the children, most of whom were under a year, with some said to be "dying" or "extremely sick." He further admitted that the children's "prolonged reaction sometimes occurs which is not pleasant to see" and often necessitates "the hands of the children [being] confined during the first twelve hours to prevent any rubbing of the eye."[76]

**IN COMING YEARS, RESEARCH ON CHILDREN** would become more frequent and varied. Whatever the scientific quest or particular medical mystery under investigation, researchers knew where to go to acquire the clinical material they needed for their studies.

For example, in 1954, Chicago-based doctors Louis W. Sauer and Winston H. Tucker used a total of "213 infants" in a series of dosing-level studies. Of that number, "132 [were] Evanston Infant Welfare children attending the Evanston Health Department Immunization Clinic," and the "remaining 81 infants [were] orphans at the St. Vincent Infant and Maternity Hospital."[77] In an effort to determine the safety and efficacy of "immunization procedures against diphtheria, tetanus, and pertussis," the doctors selected healthy children from a Chicago institution that had not suffered an outbreak of any of those diseases. Sauer was an emeritus professor of pediatrics at Northwestern University; Tucker, the commissioner of health in Evanston, Illinois. The study was paid for by a grant from the Parke, Davis Company.[78]

In 1953 Dr. Johannes Ipsen Jr. traveled to Wrentham State School in Massachusetts, an institution formerly known as a custodial facility for "feebleminded" and "defective boys." He went there to perform a series of experiments on humans after he had performed similar tetanus potency studies on mice and guinea pigs. The experiment was designed to determine if several different tetanus vaccines from different manufacturers produced different results with regard to potency and effectiveness when injected in humans. According to a subsequent journal article on the project, about 150 young men aged fifteen to twenty-five were selected to receive "tetanus toxoid" injections.[79] It is unlikely that Ipsen gave any serious thought to conducting this experiment at the Boston Latin School or Harvard University, both of which were nearby.

Interestingly, in his account of his tetanus project, Ipsen mentions that the real "cost of organizing, injecting and bleeding volunteers, can hardly be estimated. Such studies are usually a matter of opportunity and the costs were in this case rather on the low side, because all 128 individuals could be injected in a few hours. The measurable costs are those involving mice."[80] Not every researcher was willing to publicly state that human subjects were available for less cost than laboratory animals.

Most researchers were far less forthcoming about their research and their reasons for choosing one test population over another. For example, in a somewhat similar nontherapeutic exercise in 1960, "26 children between the ages of 2 and 12 years of age" were given canine "distemper virus" to see if there was a immunological relationship to the measles virus. After injecting the children with the dog virus, "no clinical reactions were observed." The authors subsequently stated in a journal article, "In the absence of a significant rise of measles antibody in distemper virus vaccinated children [they questioned] the likelihood that significant measles protection is conferred by attenuated distemper virus."[81] It is not surprising that their article made no mention of where they obtained the children for their study or whether parental consent had been granted.

**IN THE EARLY YEARS OF WORLD WAR II,** researchers at the University of Pennsylvania School of Medicine and the Children's Hospital of Pennsylvania became preoccupied with "inhalation experiments" and the "effectiveness of vaccination for the prevention of epidemic influenza." It was their hope that research in this area and experiments on "vaccinated and non-vaccinated human beings" would provide some "additional information about the value of these vaccines."[82]

In one study, an experimental influenza vaccine was administered to test subjects, who were then given repeated blood tests. After receiving intramuscular injections of two different vaccines, test subjects underwent an influenza inhalation procedure whereby they were "exposed for 4 minutes" to the vaporized fluid containing a large dose of influenza germs.

The subjects for this nontherapeutic experiment were selected from a cottage at a large state institution in New Jersey. The seventy-two test subjects ranged from six to fourteen years of age and did not have the flu. After being treated and then quarantined for a period of time, eleven of the seventy-two children came down with clinical influenza, exhibiting "flushing of the face, malaise, aching of the back, arms and legs, dryness and redness of the throat, dry cough, leucopenia [whooping cough]," and a number of other symptoms, including high temperatures.[83]

**TWENTY-FIVE YEARS LATER**, in 1967, researchers at the University of Pennsylvania and the Children's Hospital of Philadelphia, including Joseph Stokes Jr. and Robert Weibel, teamed up once again to test a live attenuated mumps virus vaccine on children at the Trendler Nursery School and the Merna Owens Home, two institutions for the mentally retarded in Pennsylvania.[84] The children, who were healthy at the time of the experiment, ranged in age from one to ten. An unexpected side effect of the experiment was that three of sixteen children exposed to the vaccine came down with "clinically apparent or suggestive parotitis," an inflammation of the parotid glands (the major salivary glands located on each side of the face); parotitis is a symptom of mumps. The researchers concluded that the cases "indicated that the A level virus was not sufficiently attenuated for the purpose of routine vaccination."

**IN 1946, DOCTORS AT THE BOWMAN GRAY** School of Medicine of Wake Forest College in North Carolina began an experiment on human cross reactions utilizing trichinella skin tests.[85] Researchers felt it was desirable to use one children's institution and one adult facility for their test populations. "An orphanage and a prison, both under close medical supervision, most nearly met the requirements."

The orphanage was the Methodist Children's Home of Winston-Salem, North Carolina, and 285 children aged six to nineteen were tested for trichinella and tuberculosis after being fed killed trichinella virus. Approximately 20 percent of the children eventually tested positive.

In another experiment hundreds of miles to the north, investigators had access to custodial facilities on both sides of the Delaware River. In addition to nine children from their own backyard at the Children's Hospital in Philadelphia, they blanketed orphanages and institutions in the area. The result of their experimental dragnet included three children from Skillman, five children from Mount Holly, and six children from New Lisbon, all in New Jersey, as well as ten children from the Homewood Orphanage in Philadelphia and eleven children from Pennhurst. Approximately half of the children in both groups developed measles from the experiment.

It goes without saying that there were educational institutions closer to the Penn campus with Children's Hospital next door than Skillman in central New Jersey and Pennhurst in southeastern Pennsylvania, but the doctors behind this study knew where certain types of medical research were off-limits and where they were condoned. Orphans at Homewood, epileptics at Skillman, and the mentally retarded at New Lisbon proved ideal as test subjects. They had become that critical intervening step between monkeys and humans. Providing subjects who were cheaper than lab animals and less problematic to deal with than adults, institutions for children played a significant role in the development of many vaccines during the last century.

# SIX

# SKIN, DIETARY, AND DENTAL STUDIES

*"The Kids in These Institutions
Are So Desperate for Affection"*

**AS A FLEDGLING DERMATOLOGY STUDENT AT THE MEDICAL**
School of the University of Pennsylvania in the early 1950s, Margaret Grey
Wood was mesmerized by her young, vibrant professor. Equal parts scholar,
showman, athlete, and accomplished raconteur, Albert Montgomery Kligman
was a medical rarity, a physician-researcher who could entertain an auditorium
filled with adoring students and wax eloquent on everything from the comedic
acts and quirky personality traits of his departmental colleagues and the best
restaurants to dine at while in Philadelphia, to the most effective way to tackle
a troubling dermatological conundrum or the best methods of protecting the
skin from sun damage. And all from a man who could churn out textbooks and
medical journal articles with the best of them. That he was a much-sought-
after clinical operative for the government and the pharmaceutical industry
only enhanced his reputation as a scientific wunderkind.

"His lectures were very interesting and well attended," recalled Wood.

Kligman was very entertaining. He could make the most serious subjects
sound humorous. I remember one day he was telling us about his research at

Vineland, a school for retarded children in South Jersey. He was doing ring-worm studies there. He'd tell us about abrading the scalp of the kids and then how he rubbed *Tinea capitis* [ringworm] into the wound to create a fungus. Some of the students in the class were taken aback by Kligman's unemo-tional account of how he used the children for his research, but most were unfazed. I don't think he realized or cared how it all sounded. He thought it was fine using these kids this way. To further underscore his point, he told us, 'The kids in these institutions are so desperate for affection, you could hit them over the head with a hammer and they would love you for it.'[1]

Wood, who would go on to a long career in dermatology and become the first female chair of Penn's dermatology department, admits to being "quite impressed by Al Kligman" upon first meeting him. But eventually her venera-tion of one of the great minds in postwar dermatology would dissipate. Like Wood, other detractors who were initially impressed by his brilliance grew alienated by his cavalier treatment of people, his self-aggrandizing personality, or his perversion of medicine's true mission.

"I believed everything he said as a student," said Dr. Paul Gross, who was in medical school in the late 1950s. "Kligman was brilliant and tremendously creative. He was a genius; one of the very few that is really creative and origi-nal. He should have had a foundation to discuss new ideas and ways of doing things."[2]

But there was another side to the medical professor who often took young, aspiring dermatology students under his wing and exposed them to the finer things in life. At a moment's notice he could turn on you and make you the brunt of jokes. But as Wood explained, retaliation was out of the question. "He was not somebody you'd want to tangle with. He had power and knew how to use it. Kligman always did what he wanted to do. In fact, he would often claim that 'superior people' did not have to abide by the rules and regulations that most people have to live by."

The State Colonies for the Feebleminded in Vineland and Woodbine, New Jersey, had a long history of hosting scientific projects. Albert M. Klig-man was neither the first nor the last microbe hunter to travel to the grim institutions in rural South Jersey, but the studies he performed there and the journal articles that followed would launch him on the most lucrative phase of his controversial career.[3]

As an expert on fungus and at the time in search of an effective fungicide, Kligman recognized that institutions for "congenital mental defectives" presented ideal locations for research. "Large numbers living under confined circumstances could be inoculated at will and the course of the disease minutely studied from its very onset." And as he frankly boasted about the children's facilities, "biopsy material was freely available."[4]

In one set of studies designed to measure severe scalp trauma and supported by a grant from the US Public Health Service, Kligman took eight children between the ages of six and ten and rubbed ringworm-infected hairs over an area of their scalp. On the opposite side of the child's head, "a corresponding area was first vigorously scraped with a dull knife until there was a copious exudation of serum mixed with blood. Ten infected hairs were then applied to this abraded area. These readily stuck to the site of deposition and were trapped in the crust which formed." In total, Kligman took either a "dull" or "blunt knife" to the scalps of dozens of retarded children and then proceeded to anoint them with ringworm. "The experimentally produced lesions were not treated." Obviously, he preferred to observe the fungus run its course in these experimental subjects. In the article in which the results were published, Kligman made no mention of gaining parental permission for the experiments or notifying the superintendent exactly what he was doing to the children.

The article also included the comments of various medical experts. Not one discussant raised an objection to the fact that the study's investigator had rubbed "millions of spores" of *Microsporum audouini* (ringworm) into the "traumatized" heads of institutionalized children. In fact, one much-respected Ivy League dermatology professor was so pleased with the experimental exercise that he commented, "The employment of human test subjects is ideal" and admitted that he, too, had once applied "fungicidal agents" to inmates at two penitentiaries. He then advised his fellow investigators, "We have not been alive enough to the wealth of test material" in the nation's custodial institutions.[5] In addition to his position at Penn, the reviewer, Frederick Deforest Weidman, was a former president of the American Dermatological Association and vice president of the American Board of Dermatology and Syphilology. His views and endorsements carried weight.

In another ringworm experiment that tested various chemical preparations as potential treatment modalities, "various strengths of formalin solution were applied to the scalps of ringworm patients." In one trial, "a tightly fitting rubber bathing cap was placed over the head of the patient and a three inch incision made into the dome of the cap. Through the hole was inserted about six four-inch gauze squares. One hundred milliliters of formalin were poured onto the gauze. The hole was sealed with adhesive." In a number of cases, the strength of the formalin—a substance similar to formaldehyde considered toxic and ultimately carcinogenic—was increased. "In six cases," according to Kligman, "the individuals were exposed to full strength formalin for one hour." Regrettably, Kligman was forced to admit, "in none of these was a cure effected." However, and almost as a surprised admission, considering the painful nature of the procedure, Kligman stated, "One child in a state mental institution was able to tolerate the formalin treatment for five hours."[6]

Kligman's reputation soared during the Cold War years, and he became something of a dermatological celebrity, known for his innovative and varied research interests, his lucrative relationships with major pharmaceutical firms, and his discovery of such popular skin creams as Retin-A and Renova. Several colleagues came to believe he had single-handedly turned a minor medical sub-specialty of "pimple-squeezers" into a profitable, respected profession. But to a certain few—his critics—Albert M. Kligman represented the worst of the medical profession: physicians who were motivated by commercial and economic interests and willing to use everything and everyone to accomplish their goals.

Regardless of one's views on Kligman's career and contributions, he is arguably one of the best representatives of an unfettered research atmosphere during the 1950s, 1960s, and 1970s that allowed medical investigators to use retarded children, geriatric patients, and prison inmates as docile, unquestioning test subjects. Scholars can debate how important eugenics, economic self-interest, and utilitarianism were in shaping Kligman's view of human research, but few can doubt his indefatigable, freewheeling approach to scientific investigation and the acquisition of "research material." As Kligman nostalgically said of the Cold War era, "Things were simpler then. Informed consent was unheard of. No one asked me what I was doing. It was a wonderful time" to do research.[7]

**ALTHOUGH KLIGMAN WOULD GO ON** to perform a wide array of skin experiments—not to mention a broad spectrum of other clinical trials—on prison inmates and indigent geriatric patients in coming years, he was certainly not the only physician-investigator to perform skin experiments on institutionalized children. He and others were just following a long-established tradition of taking advantage of the "material" in custodial institutions to test new theories on prevention and treatment as well as practice various aspects of their chosen craft.

Dr. Botho F. Felden, for instance, came up with the notion that thallium acetate, a highly toxic agent and subsequently proven carcinogen, would be an effective treatment for ringworm in the scalp of children. In an attempt to produce alopecia to combat the fungus, Felden went to the Children's Hospital on Randall's Island in New York City and targeted forty-seven children with trichophytosis (a form of fungus) of the scalp. "All of the children," according to Dr. Felden, "had various degrees of feeblemindedness. They consisted of morons, imbeciles, and idiots between 2 and 19 years." Forty-five of them were between 2 and 13 years of age.[8]

Thallium acetate's toxic impact was no secret. Those who cared to read the medical literature had long known of its potency. As Drs. Robert G. Swain and W. G. Bateman commented on the basis of their own experiments in 1909, which resulted in the deaths of a wide assortment of lab animals, including large dogs, "thallium deserves to be classed among the most toxic of the elements, progressing in its physiological action with remarkable certainty and definiteness." In fact, the authors said, it "ranks very close to arsenic."[9]

A doctor wrote of his investigations in the 1920s: "The effect of a single dose of thallium acetate is well known; about the seventh day the hair begins to loosen, and by the fourteenth day it comes out in a most dramatic fashion. By the nineteenth day depilation is complete." Though pleased with such results, the downside for patients included joint pains, loss of appetite, irritability, damage to the thyroid gland, and a serious condition involving low gastric acid called "achlorhydria." Some doctors, however, believed young children were "more tolerant of thallium" than older children. "My youngest patients have been 1 year old," boasted one physician.[10]

Apparently, Felden was pleased with his results as well, though toxic symptoms did occur in his test population. Undeterred, he considered his experiment satisfactory and informed doctors that "thallium acetate is a valuable drug in producing epilation in children." However, "The indiscriminate use of thallium by those not thoroughly familiar with its indications, contraindications and its grave toxic qualities is warned against emphatically."[11]

Years later, another physician who claimed to be concerned about healthy skin traveled to the Laurel Children's Center in Maryland and gave fifty patients with mild to severe acne the drug triacetyloleandomycin, or TriA. The center, opened in the 1920s as Forrest Haven, gradually suffered budget problems and fell into disrepair. Residents were routinely abused and neglected, and many were sought out as test subjects for medical research. One doctor who traveled to the facility was Howard Ticktin of George Washington University Medical School. In "an attempt to determine the incidence and type of hepatic dysfunction relatable to the administration of TriA," Ticktin determined that after two weeks of treatment, "more than half the patients showed hepatic dysfunction."[12] In fact, eight had to be transferred to the hospital for treatment: Six showed symptoms of anorexia and abdominal aching, and two became jaundiced. Ticktin's decision to give a "challenge dose" of the medication to four of the recovering patients and a "second challenge dose" to one subject was at the very least risky, if not wildly imprudent even by the loose medical standards of the day. However, actions against physicians were a rarity at state facilities. That is a key reason why they performed clinical trials at such institutions: There was little fear that an untoward result or serious injury would result in any legal complications.

Early in the twentieth century, for example, doctors were befuddled with the problem of spontaneous hemorrhaging or bleeding in some newborns. Medical researchers came up with some bizarre ways to remediate the problem.

One such "remedy" consisted of a gelatin solution infused subcutaneously between the shoulders as a blood thickening and packing device with some oral doses as well. In one chilling case, a doctor apparently tried to inject the gelatin into the child's rectum. The children often became very sick with toxemia-type symptoms, including fever and an increased pulse and respiration rate. The mixed results—some children died while others recovered—seemed to confirm that gelatin did coagulate the blood, but "the ineffectual use of gelatin in case 12 [where a child died] led us to do some experimental work on normal children with gelatin."[13] Those who were then subcutaneously infused with gelatin were hospitalized for various reasons, though they were not said to be bleeding. This small sample included a four-year-old "feebleminded" malnourished child, an ill nine-year-old boy, and a twenty-two-month-old. All three became quite ill from the gelatin infusions. The infant apparently suffered "alarming symptoms of prostration and collapse, with an elevation of temperature and an acceleration of pulse and respiration."[14]

Interestingly, Dr. Isaac Abt, the theorist behind this dubious project, grew defensive about the quality of the sterilized gelatin in the study and admitted, "It was not deemed wise, however, owing to the grave symptoms of toxemia which resulted, to carry on the work any longer with children, and therefore, a few rabbits were injected." Apparently the animal experiments turned out as poorly as the human trials—some died—leading Abt to believe "the explanation of the toxemia produced by the gelatin is not very far to seek, when one recalls that the gelatin is manufactured from the bones of animals." In other words, Abt believed "the decomposition which takes place in these bones gives rise to cadaveric poisons . . . and these ptomains [*sic*] are contained in the gelatin." Despite his admission that "it would be difficult to state what a safe dose of gelatin should be," he still "warmly recommend[s] the gelatin treatment for the newborn infant."[15] We cannot ascertain whether Abt tried his gelatin concoctions on animals before moving to humans—or, at least, "feebleminded" children; if he did, he proceeded to move forward long before all the data were tabulated and his therapy had been proved safe.

**SLIGHTLY MORE THAN A HUNDRED YEARS AGO**, one of the greatest plagues gripping America was pellagra. An ugly and much-feared skin disease particularly prevalent in the Deep South, it was characterized by the four Ds: dermatitis, diarrhea, dementia, and death. During the first half of the twentieth century, there were an estimated 3 million cases, with 100,000 ending in death.[16] In 1914, there were an estimated 50,000 cases, 15,000 in Mississippi alone. At least 10 percent of those resulted in death.[17]

The medical establishment was at a loss as to pellagra's origin and treatment. The dreaded disease was characterized by a red rash that covered the face, hands, and feet and turned dry and scaly over time. Victims became physically weak and easily disoriented, and eventually they suffered acute diarrhea and pronounced mental aberrations. Most doctors thought pellagra an infectious disease; many eugenicists argued that it demonstrated longstanding hereditary defects aggravated by poor personal hygiene and inadequate sanitation. Frustrated by the lack of progress, one prominent public health official from South Carolina who was wedded to ancient myths and eugenicist notions described the disease as "the greatest riddle of the medical profession," and "a sphinx of which we have asked a reply and gotten none, for nearly two hundred years."[18]

The intractable scourge finally met its match in Dr. Joseph Goldberger. A dedicated US Public Health Service official who was drawn to science, avoided myth, and disdained eugenic propaganda, Goldberger took an extended tour of the South in 1914 to better observe the disease. He visited some of the locales that appeared most victimized and specifically examined public institutions that seemed hotbeds of pellagra. His observations were both sound and revolutionary. Most important, he forcefully argued, pellagra was not a contagious disease. Institutional residents with the disease never gave it to staff members. Second, the disease was not a direct result of poverty or defective hereditary traits as eugenicists were suggesting.

So what was behind the age-old pellagra dilemma? Goldberger targeted a dietary connection. From his travels and particularly in state institutions where the disease usually ran rampant, he could not help but notice how some foodstuffs were plentiful while others were practically nonexistent. If the disease was to be stamped out, theorized Goldberger, the first and most important remedial measure was to ensure that everyone had access to a full and balanced diet. He argued for the "reduction in cereals, vegetables, and canned foods that enter to so large an extent into the dietary [sic] of many of the people in the South and an increase in the fresh animal food component such as fresh meat, eggs, milk."[19]

He admitted to a "grave doubt," however, that his curative prescription would be taken seriously without a "practical test or demonstration" for community leaders. Since pellagra did not occur in animals, human trials would be required. Goldberger knew just where to begin his studies. The Mississippi Orphans Home of the Methodist Episcopal Church South in Jackson had over 200 inmates. At any given time, a quarter to a third of the children there displayed the red rash of pellagra. Once the facility's superintendent granted permission for the experiment—based on assurances that the US Public Health Service would pay for the additional foodstuffs—Goldberger went to work. He made sure the children's regular diet was augmented with additional items, especially fresh meat, milk, and eggs.

Not only did the orphanage's residents appreciate the dietary changes, their health improved dramatically. He had indeed solved the riddle of an age-old disease, but Mississippi, a region of the country steeped in eugenics lore, resisted his conclusions and refused to implement the necessary changes. It soon became clear that the eradication of pellagra would not come overnight. Goldberger would have to repeatedly explain and demonstrate "that animal protein in the diet cured pellagra." His most famous experiment was at the Rankin Prison Farm in Mississippi, where, over months, "volunteer" prisoners had items withdrawn from their diet until the repugnant red flame of pellagra appeared across their bodies. Some nearly died in the process, but Goldberger had definitely established that the disease was of dietary origin. For all his brilliance and commitment to public health, it should be underscored that even Goldberger felt comfortable using institutionalized children and prisoners as test material to prove his theories. Enlightened and intuitive when it came to medical conundrums, he fell back on the age-old practice of using devalued populations to solve public health problems.

For the children of one particular Mississippi orphanage, however, a doctor's innovative medical experiment had proved both enjoyable and life affirming. Not all such children in custodial institutions across the country were so fortunate.

**AS CAN BE SEEN FROM EARLIER DISCUSSIONS,** doctors were not averse to incorporating even the very young in clinical trials. In Detroit in 1932, for example, researchers working on rickets turned to the city's newest and poorest members for additional knowledge about the disease. They chose three- and four-month-old infants from Detroit's Child Welfare Clinics—199 of them in total—and "all free from rickets clinically."[20] They were then divided into three groups and given various combinations of evaporated milk, pasteurized milk, and cod liver oil in an effort to determine what was most and least effective in combatting rickets. The infants received monthly X-rays of their legs, a radiation regimen that even doctors in the 1930s must have known was not recommended for young children.

A decade later, during World War II, doctors began a number of dietary experiments on youngsters that would be of no benefit to them. One such study on young males at an institution and supported by the Mead Johnson Company was designed to induce thiamine deficiency. Subjects were intentionally given a diet deficient in nutrients and vitamins and just fed granular dough three meals a day for more than eighteen months. Besides the bread-like meals, there was no other source of nourishment. Though the investigators admitted the diet was "monotonous," they told others "the subjects soon became accustomed to this food and ate it with every appearance of relish."[21] In addition to the extreme diet, test subjects were repeatedly put through blood, urine, and fecal tests, plus periodic electrocardiograms. It was no surprise when the children did develop thiamine deficiency. Thiamine deficiency can impact all organs in the body but is especially destructive of the nervous system and often leads to other health maladies, such as beriberi. In addition, the experiment resulted in children vomiting and showing signs of anorexia. Obviously pleased with their results, Victor Najjar and L. Emmett Holt of Johns Hopkins moved from thiamine to riboflavin and did a similar experiment with twelve ten- to sixteen-year-olds in an undisclosed institution for three months.[22]

Less than a decade later, during the Korean War, researchers from Texas State College and Pennsylvania State University went to three orphanages to put groups of children through a series of bread enrichment studies. The subjects ranged from six to fourteen years of age and were fed four or five slices of bread daily that contained various enrichment levels of thiamine, riboflavin, niacin, and iron. During the thirty-six-month study, the children were examined to gauge the impact of enriched bread being added to and discontinued from their diet. To the researchers, how this impacted the health of the children was important only from the standpoint of the numbers resulting from their blood, urine, and other tests.[23] Interestingly, researchers were able to document the positive effects of enriched bread on the general health of the children, but they made no great plea or effort to ensure that children at the orphanages would receive enriched bread as a regular part of their diet. In the end, the children were just subjects in a clinical trial.

In another example during the early 1960s, doctors at the Long Island Jewish Hospital were determined to establish whether lactic acid milk products produced acidosis in newborn infants. Acidosis, the presence of excessive amounts of acid in bodily fluids, including blood, urine, and tissues, can be dangerous and cause weakened body systems, a perilous dilemma for infants, especially those born prematurely. To assess this problem and satisfy their curiosity, they decided to intentionally administer lactic acid to healthy babies born prematurely to see if they did worse than other babies. Not only were various formulae given to the infants with the likelihood of producing acidosis, but infants also had their femoral veins punctured for blood analysis. Over seven days, eighteen infants were fed lactic acid–enriched evaporated milk.

Doctors then examined how acidosis impacted the infants' carbon dioxide, plasma, blood, and lactate levels. They also injected four of the infants with sodium lactate to determine if their blood lactate levels increased. In fact, their levels did increase. The infants proved vulnerable to acidosis. In the words of the authors, "many were not capable of handling this relatively small acid load and developed acidosis."[24] Such studies were occasionally replicated, each one putting newborns—some as young as two days old—at risk. The surprising conclusion of all these studies: "It would seem advisable that lactic acid milks should not be used in the nursery for premature infants."[25] Moreover, the article never mentioned anything regarding parental permission, a dubious prospect at best if parents understood what their babies were going to endure.

Putting infants through invasive clinical procedures to acquire new information was far from an unusual practice. Adding to the corpus of medical knowledge was a respected tenet of the academy. Attaching their names to articles in respected medical journals benefited researchers in a number of ways. The rewards—both professional and personal—far outweighed the potential harm that might be inflicted on the study's subjects. For example, the authors of a gastric emptying study that voided the stomachs of subjects in the early 1960s gave various solutions, including barium sulfate, to 148 premature infants so that abdominal X-rays could be taken.[26] The experiment resulted in a journal article and increased knowledge of the gastric tract, but did the twelve dozen infants benefit from having their systems flushed with chemicals and then being forced to endure unnecessary X-rays?

**THE COLD WAR ERA WAS ALSO** a period of intense experimentation for the dental profession. Dentists initiated numerous studies on everything from the effect of fluoride-containing toothpaste on tooth enamel to the impact of "refined sugar" on cavity development. Once again, and consistent with the practice of their medical colleagues, a disproportionate share of these studies was undertaken at orphanages and schools for the "feebleminded."[27]

One 1951 study "recognized the need for a well-controlled, long-term, individualized study of caries progression in children's teeth under conditions which would permit control of as many variables as possible."[28] When the opportunity allowed, admitted the authors, they jumped at the chance "to use subjects already living under conditions of limited regimentation. The most appropriate group for our purpose was that to be found in one of our state schools for the mentally retarded."[29]

The study, which was sponsored by the Sugar Research Council, included 200 children between the ages of thirteen and twenty whom the researchers believed had the "capacities for adequate cooperation" and would not be particularly difficult as test subjects. To ensure that they would disrupt institutional operations as little as possible, the doctors admitted that the "subjects had received only minimal amounts of dental repair prior to the beginning of our study, and during its course only emergency reparative services were given. Cavities initially present remained untreated and unfilled throughout the course of the study."[30] Interestingly, the authors found the diet of the children "so deficient" that they had to alter it dramatically. While one group avoided all refined sugar, the other was "provided not less than three ounces of sucrose daily, usually as a constituent of foods, sometimes as candy which would be eaten at the table with the meal." The result of the sugar industry's two-year study found "no significant difference" among adolescent children when the ingestion of refined sugar was arbitrarily prohibited or given daily in large amounts.

Other dental studies from the early postwar period through the 1980s continued to use "regimented" institutions that provided cheap and easy access to their inmates. Dental studies measuring everything from sweetened versus unsweetened ready-to-eat cereals to chewing gum were commonplace during these years. Articles on experiments that noted parental permission and noninstitutional confines tended to report studies that reduced cavities and gave children a choice of foods. Those reporting on experiments in custodial institutions and without parent involvement discussed studies that risked producing cavities and offered a far less attractive diet to the participants.[31] The articles also occasionally disclose corporate sponsors, such as General Mills in the cereal study.

An example of a study that harmed children occurred in 1972 at the Lincoln State School in Illinois, where 567 institutionalized children aged eight and older with an IQ of 20 or more were incorporated into a study to determine the effect of carbonated beverages on the incidence of cavities.[32] The Lincoln State School and its mentally retarded residents had been merged with the Illinois State Colony of Epileptics to form an inviting, if not irresistible, pool of potential test subjects. In his acknowledgment section of a subsequent article in the *Journal of the American Dental Association*, Dr. A. Steinberg, a periodontology professor at the University of Illinois College of Dentistry, thanked the superintendents of the Lincoln State School, the Illinois Soldiers and Sailors Children's School, and the Glenwood School for Boys for their cooperation in allowing studies at their institutions.

Whether dermatologists, dentists, or others in the medical profession, researchers knew where to go to conduct clinical trials. No exact number can be attached to these studies, but we believe hundreds—if not thousands—of children over the years were forced to become subjects for scientific investigation.

# SEVEN

# RADIATION EXPERIMENTS ON CHILDREN

## "The Littlest Dose of Radiation Possible"

*A stinking Mickey Mouse watch wasn't worth all they put me through.*

—Charles Dyer

WHEN GORDON SHATTUCK WAS BROUGHT DOWN WITH AP-proximately twenty other children early one morning in 1950 for a special meeting in the boys' dormitory, there was an air of excitement and expectation in the room. Dr. Clemens Benda, the medical director of the Walter E. Fernald School, was there along with several other people. Most of the adults present were unfamiliar to the boys, but they all looked official and important and referred to each other as doctor or professor. They had smiles on their faces and seemed eager to talk to the boys. The guests, from the Massachusetts Institute of Technology (MIT), were about to offer the kids at Fernald a rare treat, an opportunity few would turn down.

"This was something new and really out of the ordinary," recalls Shattuck, who was barely twelve at the time. "Most of us never had visitors, not even our parents came to see us, and to be brought down for a meeting with strangers was really unusual. It was exciting. They told us they were starting a new club, a special Science Club that would be involved with some interesting experiments."[1]

A curious and handsome boy who displayed natural leadership qualities, but not always for the best of endeavors, Gordon Shattuck had difficulty containing his enthusiasm for the new program.

Shattuck, who is now in his mid-seventies, doesn't recall much being said about the scientific experiments the doctors had in mind at that initial meeting, but he and all the other boys clearly remember mention of special events and fun-filled excursions. "We all got excited," says Shattuck. "We never got out of Fernald; we never left the institution. It was terrible in there, and now they were offering us trips to Fenway Park to see the Boston Red Sox and the Boston Braves and to go to MIT for parties and tours of the college. Everybody agreed, it sounded great. I remember them saying it was going to be something like the 4-H Club. We all thought it was wonderful."

It would not remain wonderful for long. Fairly quickly, Shattuck and many of his friends in the Science Club would come to loathe the scientific studies they were now part of, and the few incentives they were being tossed did not come close to ameliorating their fear of the project. That feeling would only intensify over time, especially when they finally learned the truth about the Fernald medical experiments.

The Shattuck household in Marblehead, Massachusetts, was a crowded, raucous, generally challenging environment. There, one did not grow up as much as struggle to survive. Though it was good preparation for what Gordon and several of his siblings eventually would have to endure at Fernald and other institutions, there can be little doubt that the children would have much preferred to have grown up in a healthy, loving home.

Gordon was born on September 9, 1937, to Henrietta and Gordon C. Shattuck Jr. His father was a big, intimidating, hard-drinking man who ruled his ever-growing roost in ruthless fashion and seemingly did his utmost to make everyone there miserable. A six-foot four-inch, 250-pound short-order cook who worked in a series of Massachusetts diners when he was sober enough to do so, he was cursed with an addiction to alcohol, a terrible temper, and little regard for his wife and children. When he wasn't spending his modest wages on beer and liquor or pummeling his wife in recurring and violent domestic disputes, he managed to father a staggering twenty-two children.

Gordon's mother—a besieged and occasionally bloodied housewife with her own alcohol problems—was a baby factory. As the oldest boy in a rapidly growing and unruly household, Gordon felt neglected and abused and was ever in search of parental affection.

"My father beat me all the time," Gordon says matter-of-factly. The few good times are hard to remember and pale in comparison to the many bad times, especially when he witnessed his mother getting physically abused. "He'd come home drunk, they'd argue, and then he'd beat her up." Eventually, she would succumb to the pummeling and pervasive hopelessness and try to take her own life. Gordon, a young child at the time, watched in horror as his mother took a kitchen knife to her own throat.

Shattuck recalls, "Blood was on her, on the floor, it was everywhere. It was terrible. She nearly died. They had to give her five transfusions."

It wasn't long after that incident that Gordon was first taken away from his family. The state determined that the Shattuck home was a dysfunctional, dangerous environment and no place for young children to be raised. "State ladies came," recalls Gordon, "and took six kids away. I was about five years old and went to a series of foster homes. I hated them. There was only one that was any good, most were terrible, and I often ran away. Then they would capture me and I'd be returned to the same foster family or some other people. I must have been in a hundred homes over the years and ran away as often as I could. I missed my brothers and sisters."

After many foster homes and a few institutions such as Boys Haven, a Catholic shelter for troubled boys, the Massachusetts Division of Child Guardianship decided that some of the Shattuck children needed a more rigid and structured environment. Brother Bobby and sister Dottie were sent to Wrentham State School, and Gordon was sent to Fernald.

When Gordon entered Fernald in the spring of 1947, he was just nine years old. Frightened, lonely, and desperately missing his siblings, he was about to embark on as painful and troubling a journey as one can imagine a young boy taking in postwar America. Over the next half-dozen years, Gordon Shattuck would be subjected to constant verbal and physical abuse, repeated sexual assaults, and extended periods of forced labor. He contemplated escape on a daily basis.

Fernald was a large state institution housing nearly 2,000 children and adults with a stunning array of afflictions and disabilities. The size of the place

and the many severely impaired people warehoused there alarmed and frightened the boy. The institution struck him as a combination of prison and human zoo, and it made him miss his mother and siblings all the more. Initially,
he underwent a battery of physical examinations, interviews, and tests. Documents were collected from his family and various state agencies. The papers
disclosed much about his life. He had weighed "9 pounds at birth . . . began
to walk at 18 months . . . and talk at 1 year." He had all the normal childhood
diseases, such as whooping cough, measles, chicken pox, and German measles,
and he was declared healthy and sound of body, but his psychological state and
educational status were other matters.

His "peculiarity" was first noticed "when he was unable to compete in
public school." Gordon spent two years in the first grade and shortly thereafter was "placed in a special class where he accomplished very little," and his
classroom effort was described as "poor." He was further said to be "egotistical,
selfish," and prone to "run away from foster homes if he is not occupied" with
something he enjoys doing. It was recommended that he needed direction
and a "firm hand."[2] The leadership of Fernald planned on giving him that and
much more.

Walter E. Fernald, the school's third superintendent, was an internation-
ally recognized expert on mental illness and mental retardation and placed
the institution firmly on the educational map. After Fernald's death in 1924,
the institution would adopt his name, maintain its allegiance to a eugenics
movement and philosophy that dominated a good portion of the intellectual
arena in the early decades of the twentieth century, and become home to a
cross-section of people with developmental disabilities.[3] It would also increasingly become home to those without physical and psychological disabilities:
orphans and other children from poor, broken, or dysfunctional homes, delinquents, and epileptics. Even those who just scored low on IQ tests found
their way to Fernald. Many of this group would become part of the free labor
pool that sustained the institution when state funding did not keep pace with
its ever-growing population and responsibilities. As Ransom Greene, Walter
Fernald's successor as superintendent, readily admitted regarding his financial
constraints, he "needed a mix of 30% morons to keep the school operating."[4]
"The upper level of mental defect," Greene went on to explain, "helps in the
care of those less able to care for themselves, or we would have a very much
larger employee roster."[5]

Gordon, and many boys like him during the 1940s, 1950s, and 1960s such as his friends Charlie Dyer and Austin LaRocque, would become the backbone of Fernald's "inmate labor" system and representative of the "powerful financial incentive" to keep relatively normal children on the premises. Stigmatized with the classification of "moron" after achieving a 74 in his Fernald entrance examination, Gordon and other State Boys, as they were known, with below-average IQ tests became the mid-twentieth-century version of indentured servants, nonpaid laborers critical to the institution's upkeep and performance. Every new "moron" sent to Fernald meant one less painter, one less postal messenger, and one less gardener overseers needed to hire.

Case records of children at Fernald meticulously charted the results of medical examinations and changes in each child's behavior. They even made note of masturbation rates, weight gain, and those with whom each child associated. Gordon, for example, was described as a "pleasant, serious, attractive little boy" who was in "good health," and competent at "braid weaving and brush weaving," skills that every Fernald resident was expected to learn. As the months and years passed, however, assessments of his interests and moods would take on a darker cast. He would be described as "very stubborn and sly," possessing a "very mean disposition," and "an inveterate liar" who thought "himself better than the other boys." His "habits," such a "running away," getting into "fights," and exhibiting a "saucy and defiant" attitude, were regularly noted. The fact that the others boys held him in high regard—"hero worship" in the opinion of administrators—was of increasing concern.

Interestingly, for all the scrupulous attention to Gordon's "sneaky" ways, "troublesome" attitude, and displays of "antagonistic" behavior, there is no mention in his case folder of the harmful impact Fernald was having on him. The school's personnel and unforgiving practices were taking their toll. Sexual assaults by staff were practically an everyday—or, more accurately—an every night occurrence for him and other boys. "I don't know how many times I was raped," Shattuck admits. "They'd grab you out of bed at night and take you to the dayroom. Sometimes they'd bribe you with candy while other times they'd beat the shit out of you if you didn't do what they wanted."

Gordon says attendants—many of whom turned out to be sex offenders— would pick out one of the thirty-six boys in the dormitory, take him to an unoccupied room, and then sexually assault him. "I was beat up really bad one night," recalls Gordon. "I didn't do what he wanted. It was in the winter and he opened all the windows, made me take off my nightgown, and threw a bucket of cold water on me." Gordon says he and the other boys told Fernald officials about the assaults, "but they wouldn't believe me. We were considered idiots, morons, and retards. No one listened to us. They knew it was going on, but they didn't do anything about it. That's why I would run away. I hated that place."[6]

In addition to the sexual assaults, Gordon and most of the others designated as "morons" were forced to work in an assortment of physically demanding jobs that grown men usually did in the free world. "They put me on a weaving machine when I was nine years old," says Gordon. "I'd work on it four, five hours at a time, sometimes all day. In the summers I worked on the Fernald farm planting and picking crops. We'd be out in the hot sun a full day. I wasn't much more than nine years old, and I was working from eight in the morning until five in the afternoon. Then they'd march us back to our dormitory as we held each other's hand and then they'd feed us."

The boys were treated like unpaid field hands during the growing season and performed other difficult tasks the rest of the year. Charlie Dyer was thrown in the kitchen and scullery at the age of twelve and informed he was going to be a meat cutter. Before that he was a mattress maker and also spent time running a weaving machine.[7] Austin LaRocque was assigned the job of mailman and spent his days going from building to building delivering letters, packages, and messages. The school had a large library, but boys like Gordon, Charlie, and Austin were never taught to read. Even if they had mastered the language, it is unlikely they would have been interested in reading works that Fernald had in abundance—titles such as *Eugenics, Safe Counsel or Practical Eugenics*, and *The New Eugenics*.

If the sexual assaults, work assignments more reflective of the nineteenth-century convict labor system, and generally abysmal living conditions weren't bad enough, the institution had one more thing in store for its young, devalued residents: participation in the exciting new Science Club.

Initially, Gordon was thrilled with the notion of the club. Only a handful of special students were selected, and it seemed like a great opportunity to join the real world. "It sounded great," recalls Gordon, "we all got excited. It was a chance to get out of the institution."

Austin LaRocque is in total agreement. "I spent five years looking out a window waiting for somebody to visit me; waiting for somebody to come and take me on a trip. You wouldn't believe how important that was to us at the time. When you're in a place like that you respond to anything that you perceive to be to your benefit. Trips outside the institution were by all means a real inducement."[8]

Their enthusiasm wouldn't last long, however.

"They put us in a separate unit and we were no longer allowed to associate with the other kids," says Gordon. "They kept us apart from everyone else and made us eat special foods. They made sure we ate everything on the plate. They gave us oatmeal for breakfast every morning and made sure we ate all of it. Worse yet were the needles. I hated the needles. Every morning before breakfast they'd give us a needle, take a vial of blood, and make us piss and crap in glass jars while nurses watched us. I really hated it."[9]

The boys enjoyed the occasional outings but became increasingly petrified of the needles, an everyday occurrence. Regardless of whether the doctors were extracting blood or injecting them with some unknown concoction, the entire exercise had lost its appeal. Gordon finally told the doctors and nurses he had had enough; he wanted out of the Science Club.

To Gordon's alarm, he would not be allowed to drop out of the program. "I refused to take another blood test," he says. "They told me if I didn't let them take blood, they were gonna put me in seclusion. I said I didn't care, I didn't want any more needles."

True to their word, the doctors and Fernald administrators put Gordon in Ward 22, a building with six special punishment cells in the basement known to Fernald residents as the jail. "They put me in a little room with bare walls and an old ratty mattress and a can to piss in," says Gordon. "There were no windows except for a little pane of glass in the door so people could look in and check to see what you were doing. For the first few days they just gave me bread and water. Then they came and asked me if I was ready to rejoin the Science Club and I told them no. I wasn't going back; I didn't want any more needles."

Gordon spent several more days in seclusion. By the eighth day, however, he had been worn down. When the authorities again asked if he was willing to rejoin the program, he conceded. The doctors sweetened the pot by promising to take him and the other members of the club to a baseball game.

"I didn't want to do it," Gordon says regretfully of his decision, "but I said okay. I was just a little kid, what could I do?" Study organizers may have downplayed the burden on participants, but children quickly recognized the time, isolation, and discomfort associated with it. Documents eventually uncovered showed that a test subject had to endure six blood samples and four urine samples between early morning and two-thirty in the afternoon.[10]

Gordon Shattuck wasn't the only State Boy to demonstrate defiance regarding participation in the Science Club. There were others who lost their enthusiasm for Fernald's version of the 4-H Club. One of these boys was Gordon's friend Charlie Dyer. Also with a below-average IQ and from a broken home, he was just entering his teenage years, and the new club seemed like a welcome diversion from the tedium and routine of institutional life. "I jumped at the chance to get in the Science Club," Dyer readily admits. "I would do anything to get out of Fernald, even for a day."

Rather quickly, however, his interest in the program waned, and it soon became something he feared and despised. Like Gordon, he wanted to leave the club. He told the doctors and administrators that he wanted to return to his original ward and end his association with the Science Club. Dyer was stunned and angered when told he couldn't withdraw; he had signed on and would have to remain involved until the program concluded. The next time the doctors and nurses went to take his blood, Charlie wouldn't let them. They, in turn, wouldn't let him leave the building and continued to chase after him for a blood draw. The battle of wills resulted in something no one at Fernald had ever seen before: Dyer climbed up a support column toward the ceiling and shimmied across a crossbeam high above the floor. Doctors, nurses, school attendants, and institutional residents pleaded with him to come down, but Dyer refused. He was determined not to go back to the Science Club.

Dyer said that even though he was scared of heights, he "climbed up there to get away from those needles."[11]

The standoff lasted for hours, and with each passing hour, the staff grew more frantic. Visions of a twelve-year-old boy falling thirty feet to his death shot through everyone's mind, but for all their threats, promises, and words of encouragement, Charlie refused to climb down.[12] Finally, after four excruciating hours, Dr. Malcolm Farrell, then superintendent of Fernald, was brought in and talked Charlie Dyer down. And just like Gordon Shattuck, he was forced to remain in the Science Club, eat his morning oatmeal, give blood, and continue to do his business in glass jars.

Eventually, many of the State Boys would be able to put both the Science Club and the Fernald Training School behind them. After their release from the institution—usually in their late teens and early twenties—they made their way in life by getting jobs and earning a living, getting married and having children, and doing the best they could after being dealt a bad hand by their parents and the state. Recognizing that they had lost any semblance of a normal childhood and bore the stigma of being officially declared morons wasn't easy, nor was the challenge of going through life unable to read, but many persevered, and their struggles paid off. They became contributing members of society, and few who met them could ever imagine the horrors they had endured. The acclimation process was far from easy, however. The famous century-old state institution had not prepared its graduates to enter society. "When I finally got out of there at the age of twenty," says Charlie Dyer, "I couldn't read and had a hard time functioning. They never taught me anything. I didn't even know what money was. I had worked all my life but never got paid for anything." He was not alone. "I didn't know what to do," says Austin LaRocque of his release from Fernald. "I never rode on a bus before. I didn't know how to use money. I didn't know what money was. I was totally unprepared to make my way in the world."[13]

But Charlie, Austin, and Gordon did make their way in the world only to receive one last shocking blow, a blow that harked back to their days in the institution in Waltham. On the day after Christmas in 1993, the *Boston Globe* published a front-page story that shocked local citizens and reverberated throughout the country. The banner headline, "Radiation Used on Retarded," and the subtitle, "Postwar Experiments Done at Fernald School," would jump-start a media frenzy that would last many months, launch both state and federal investigations into the origin and reasons behind the experiments, and spark a national debate about medical research in America.[14]

"In the name of science," the lengthy article began, "researchers from MIT and Harvard University [had] retarded teenage boys" at the Fernald School "eat cereal mixed with radioactive milk for breakfast or digest a series of iron supplements that gave them the radiation-equivalent of at least 50 chest X-rays." News of the studies, which were funded by Quaker Oats and the US Atomic Energy Commission, sparked headlines, widespread debate, and a good bit of moral outrage. Some argued that the experiments exposed the true "face of evil," people in authority who were "infected by opportunism and arrogance" and cavalierly exploited defenseless children. Others claimed that the experiments "yielded important information about nutrition" and that no one was harmed "because the radiation levels were so low."[15]

Almost immediately the State Boys themselves were inundated with media requests. The former neglected New England waifs were now media celebrities. "I started getting calls and visits from the *Boston Herald* and the *Globe*," said Gordon Shattuck. "All the North Shore papers wanted to talk to me as well as the radio and TV stations. Even *People* magazine wanted to talk to me. They all came to my house and asked what I thought about the revelations now coming out concerning secret medical experiments at Fernald. I told them it was pretty shocking and I was beginning to wonder how long I had to live. That stuff is in your body, you don't know what it can do."[16]

Public interest would continue to build as the media became preoccupied with "retarded boys" being used as "human guinea pigs" in dangerous "radiation tests." During the coming days and weeks, investigative stories, editorials, and lengthy op-ed articles on the Fernald story would become de rigueur for newsgathering agencies across the country.[17] The *Boston Globe* underscored the inexcusable "wrongness and arrogance" of MIT and Harvard—two of the most powerful educational institutions in the country—using "retarded children" for a series of questionable medical experiments.[18]

Just a couple months earlier the *Albuquerque Tribune* had published revelations concerning eighteen unwitting hospital patients who were injected with plutonium in the late 1940s. With this new report, relatives of the victims as well as average citizens became aware of alarming stories concerning strange Cold War scientific experiments and official investigations that were now being initiated to discover the truth.[19] One such investigation was ordered by Phillip Campbell, the commissioner of the Massachusetts Department of Mental Health, who admitted to his "shock and horror" on reading the initial *Boston Globe* story of the Fernald experiments. He immediately decided to establish a task force to "fully disclose and research activity at any [DMR] facility that involved the use of radioactive substances."[20]

The fifteen-member Task Force on Human Subject Research, which included lawyers, academics, and a member of Congress as well as two former members of the Fernald Science Club, set up an 800 telephone line to enable individuals to request information, comb documents and archival material in Fernald's extensive library, and seek and receive the full cooperation of Harvard and MIT in their search for relevant information. In its effort to answer the question "How could it have happened?" the task force examined the history of guardianship and wards of the state during the middle decades of the twentieth century and the development of the concept of informed consent during those same years. It tried to build a narrative of the Fernald experience from the actual individuals who lived there, the residents themselves.[21] The narrative proved both illuminating and depressing as former Fernald residents spoke of emotions running the gamut from disillusionment to moral outrage. Charles Dyer, for example, recounted for the investigative committee that when he first entered the school, he thought he'd "learn and maybe better myself." But as the days and weeks turned to months and years, "with little understanding" as to why he was there, he realized that "it wasn't a school for learning but in fact an institution for the mentally retarded." He said he witnessed abuse routinely perpetrated on residents and provided numerous examples, such as "an employee burning a mentally retarded child with a lit cigarette." The experience did not leave him unaffected. "My bitterness then turned to rebellion," Dyer admitted.[22]

In fact, for the vast majority of Fernald residents—particularly those who were profoundly disabled and intellectually challenged—life was much worse than it was for the Science Club boys. Neglect and abuse were a way of life, and graphic eyewitness accounts of the institution's appalling practices rightfully disturbed members of the task force. One distraught mother who took her severely retarded five-year-old daughter to Fernald in 1947 spoke of her great regret "and wonder how I ever parted with my only child."

She told of her daughter being so heavily medicated that she couldn't hold her head up, "the bare rooms with cement floors" that were routinely "hosed down to remove the urine and feces," shower stalls on the outside of buildings where children showered, and the overcrowding that resulted in "very large ward-like rooms with many beds only inches apart." More troubling by far were her daughter's recurring injuries: "skin torn off both sides of her face, neck and back"; being found "bloody each morning"; and a "head wound" that resulted in a skull depression requiring several stitches without the mother being offered an explanation of what had happened.

Another guardian spoke of his care for a young girl who suffered from Down syndrome and was placed in Fernald in the mid-1960s. One of the many indignities, injuries, and indifferent care the young girl was subjected to was the decision of the institution's medical staff—without informed consent—to "extract all her teeth" so she would "not be bothered with tooth decay and cavities." For the rest of the girl's life, she would be forced to "eat pureed or ground food."[23]

Commission members heard many additional stories, such as one of "the girl found at the bottom of the pool because someone neglected to take a head count; the girl who wandered out of her building and died in a culvert from hypothermia; and the girl who received blows to the head from a male care-taker" because she soiled herself and then smeared her feces on the school's floors and walls.[24] Such personal stories underscored the miserable living conditions, overall neglect, and dubious decision-making at Fernald and may have provided the answer as to why Harvard and MIT chose to do their experiments there.

**THE ASSOCIATION BETWEEN FERNALD** and the two elite Boston academic institutions began just after World War II, when the National Biochemistry Laboratories at MIT sent a detailed letter to Dr. Malcolm Farrell, superintendent of the state training school in Waltham. In the letter, Robert S. Harris, an associate professor of nutritional biochemistry, said they were interested in conducting a clinical study at Fernald on the role of "phytates upon mineral metabolism."[25] Phytates are stored forms of phosphorous found in grains, seeds, nuts, and many cereals, and "considerable disagreement" had arisen regarding their impact on nutrition. Harris explained that some nutrition experts were advising against the use of large amounts of cereal products in the diet on the grounds that the phytates in them interfered with calcium metabolism. Harris wanted to establish the effect of phytates on iron metabolism.

He explained that earlier measuring devices were too crude for such delicate and precise scientific detective work, but the "recent production of radioactive materials by cyclotron bombardment offers a new tool for the study of materials." It would now be possible "to tag iron atoms, combine the iron into a compound, feed it and follow its metabolism in the body."

Recognizing the potential harm of radioactive material as well as the heightened concern about such matters just a few short months after the atomic bomb blasts in Hiroshima and Nagasaki, Harris assured Farrell that "very minute quantities of radioactive substance are required, and the radioactive emanation from this substance is also extremely small." In fact, it would be no more harmful than "the cosmic rays continually bombarding people living at the altitude of Denver, Colorado," he explained. To further underscore that there was nothing to worry about regarding the quantities of radioactive substances the scientists planned to use, Harris cited experiments involving "more than 100 medical students" that had proved harmless over a two-year study period and similar results in "clinical investigations at Rochester Medical School" where much greater quantities of radioactive compounds were being used.

Harris explained that they had planned to use rats in their experiments, but rodents differed radically from humans in the way they metabolized iron and phytates. Researchers thus believed that it would be necessary to use humans for the experiments and preferably subjects during a period of active growth. Additional constraints, such as the need for the subjects to remain in the study for its entirety and the need for a uniform diet, provided further advantages of Fernald as a test site. Its proximity to MIT's labs, where the blood would be analyzed, was also in its favor.

Harris then described each of the five experiments that would occur over an eight-week period and the need for ten or fifteen subjects. He also mentioned in passing that if there were any way in which they could reward the subjects, they would be glad to do so.

Over the next several weeks and into the early days of 1946, letters were exchanged between MIT, Fernald, the Massachusetts Department of Health, and the latter's Committee on Psychiatric Education and Research, which had oversight responsibilities for such a venture. Clifton T. Perkins, the department commissioner, had already signed off on the scientific exercise, stating, "I see no objections to your selecting the fifteen or so patients" in preparation for the research project, which would start in early February 1946.[26]

By midsummer, MIT scientists had concluded that phytates impair iron absorption and that this interference is much greater when phytates are administered in solutions as opposed to food. The data had been collected using eighteen children between the ages of ten and fourteen who were fed equal amounts of iron at breakfast. It was also mentioned that their milk, oatmeal, and water were "laced with radioactive material."[27]

Harris went back to Fernald in 1949 and entered into discussions with Dr. Clemens E. Benda, the institution's new medical director, requesting permission to replicate the exercise, this time doing an experiment on radioactive calcium metabolism. As with the earlier study, Harris explained that metabolic investigations of this sort were "difficult to conduct on human subjects," but

that radioactive calcium now made it "possible to obtain information on calcium retention promptly at little inconvenience to the subject."[28]

Harris explained that "radioactive calcium (Ca 45) will be administered
to adolescent children" in their oatmeal and milk at breakfast. He believed
"at least 20 normal subjects will be necessary," but there was "no limit on the
number of abnormal subjects that may be studied." By "abnormal," he specifically meant a class of children and adults who were in abundance at Fernald:
"cretins, mongoloids," and others with rare developmental maladies whose
calcium metabolism was believed to be abnormal.[29]

Each test meal would contain the minimum amount of radioactive calcium required to test the hypothesis. Each subject would participate in two
test periods, possibly more.

Consistent with gaining authorization from the Atomic Energy Commission (AEC) to acquire radioactive material, the Walter E. Fernald State
School established an Isotopes Committee to oversee handling of the material
and the nature of the research. The five-member committee consisted of five
physicians, four of them employed at the school. The fifth and lone outsider
was an instructor at Harvard Medical School.[30] Apparently no one on the
committee had concerns about the nature of the experiment, who the subjects
would be, or the level of radioactive material the subjects would receive.

Without question, the one person on the committee who should have had
the most concern about the nature and potential health impact of the proposed
MIT study was Fernald's medical director, Clemens Ernst Benda. The son of
a German pathologist, Benda was born in Berlin and studied philosophy and
medicine at the Universities of Berlin, Jena, and Heidelberg before receiving
his medical degree in 1922. He spent his early years as a medical practitioner
in clinics in Berlin and Heidelberg. In 1935, Benda and his wife, also a physician, along with their two young sons, immigrated to America because of
Hitler's rise and growing political instability in Germany.[31]

Benda settled in Boston and within a year became director of the Wallace Research Laboratory for the Study of Mental Deficiency at the Wrentham State School. In 1947 he would become the director of research and
clinical psychiatry at Fernald and would remain there until his retirement in
1963. During his career, Benda held a number of academic appointments
at respected institutions, including Harvard Medical School, Clark University, Massachusetts General Hospital, Tufts Medical School, and Boston

University School of Theology. He would also serve as president of both the American Association of Neuropathologists and the American Academy of Mental Retardation. Benda's research interests focused on Down syndrome (commonly known as mongolism at the time), cretinism, mental retardation, neuropathology, and existential psychology and psychiatry.

During their time at Fernald and for many years afterward, the State Boys would speculate about Benda's myriad research projects and recount stories of strange experiments, secret autopsies, and unmarked graves in cemeteries filled with former Fernald patients. Lurid anecdotal accounts of dozens of dissected brains in glass jars and of Benda in photographs with Nazi uniformed officers would only increase after the 1993 Science Club revelations.[32]

Though Benda and his colleagues quickly approved the MIT protocol and saw little to be concerned about, the AEC sought some revisions. The AEC's Subcommittee on Human Applications specifically stipulated "that only one dose of Ca 45 be administered to each normal child used in the study" and "no subject will receive more than one microcurie"[33]; the study had proposed two microcuries, which was equal to two grams of radium isotope. S. Allan Lough, the chief of the Radioisotopes Branch of the Isotopes Division, passed on the news of that restriction to Robley Evans of MIT and Clemens Benda of Fernald. The radioactive material was dangerous, and exposure to too great a level of the material could harm "normal children." Those deemed abnormal did not warrant the same level of concern, and restrictions on their involvement in the studies were loosened.

Correspondence between the AEC and Benda during the fall of 1949 would underscore that point as well as illuminate the inexperience of Fernald's medical personnel in handling radioactive substances. Lough would be forced to emphasize that normal patients would receive radiocalcium "one time only."[34]

Lough stated that the Fernald experiments should use the "minimum amount of Ca 45 which would provide significant scientific data in each test" and that no subject should be administered more than "one microcurie" of the radioactive material. The only leeway Lough was willing to grant concerned the number of tests in which individual subjects could participate. "Normal control subjects" could "be used in one test only with the minimum possible Ca 45." Others, meaning "mentally deficient subjects" according to Lough, could "be used in more than one test" as long as the other guidelines were followed.[35]

The level of radiation exposure each child would receive, Robley Evans assured Dr. Paul Aberhold of the AEC, could be accomplished "without going above trivial dosage levels." He then explained four simple approaches to the experiment to ensure that the doses would be harmless to subjects.[36] Evans's letter of support helped win AEC approval for the project and begin the next phase of the operation. Eventually, seventeen children would take part in the iron study.

Letters that went out to parents from Superintendent Malcolm J. Farrell notifying them of the forthcoming MIT research study emphasized that the Fernald School was interested in nutrition and that only some of the brighter patients would be participating in the project. Farrell went on to underscore the enriched diet that study participants would receive and noted that each child would exhibit "gains in weight and other improvements, particularly in the blood." He concluded his pitch for parental permission by stating: "Rest assured that I personally feel this project will be of great importance [and] eventually will be of considerable benefit to mankind."[37]

As time passed and the human research program grew to become an established part of Fernald operations and culture, additional experimental requests would be submitted and granted. In the spring of 1953, for example, Harris would inform Benda that the Department of Food Technology "planned to conduct a study of five different calcium compounds in human subjects" and requested access to "fifteen subjects." Apparently ten subjects had already been secured, but thorny problems still needed to be resolved. Three of the subjects "objected to be [*sic*] included in the study," and five others had already been used in experiments and dosed with radioactive calcium, so they could not be utilized. Harris and his medical colleagues were seeking "fresh subjects" for tests in which "each subject [would] receive 1 microcurie on each of five experimental tests."[38]

Both the request and the protocol are troubling. Just three years earlier, the AEC had stipulated that no subject should receive more than one microcurie and that only the mentally deficient could participate in more than one test. Either the rules had changed—and no such written record can be found—or physicians were taking liberties with the protocol.

Harris went on to impress upon Benda the importance of the study and his hopes that test subjects could be obtained. He appealed to Benda, asking that the "three subjects who objected to being included in the study can be induced to change their minds."[39] Such inducements for vulnerable children meant verbal threats and institutional pressure. If that didn't work, punishment—doing time in the Fernald "jail," isolation from their friends, and a bread-and-water diet—was imposed until they agreed to join the program.

The fact that subjects were opting out and boys now needed to be induced to participate reminded Harris that they had "neglected the Fernald Science Club." Harris suggested a baseball game excursion and a follow-up assembly to explain the doctors' work to get the boys "to feel satisfied their small pain is really worthwhile."[40] Unfortunately, many of the boys now viewed an increasingly rare trip to the ballpark as too modest a reward for the onerous burden of participating in the Science Club.

It is an open question as to how many of the boys' parents actually knew about the experiments; fifty-seven subjects participated in the calcium studies, most taking the radioisotopes orally while several received injections. According to Gordon Shattuck, for example, his parents were never informed and were unaware of his involvement in a science club or his being used as an experimental test subject. Nor was his name on documents representing those whose parents had signed off on their child's participation in the project. Documents do show, however, that some parents were sent notices from Benda informing them that the "nutritional department of the Massachusetts Institute of Technology" was performing "some examinations" designed to "improve the nutrition of our children."[41]

Though the one-page letter cast a positive spin on the enterprise, in actuality, the test was nontherapeutic and would provide no benefit to the boys in the study. Benda completely omitted some crucial aspects of the study, such as the use of radioisotopes, and altered other aspects to make the protocol appear either less threatening or directly beneficial to the boys. For example, according to the letter, "volunteers" would give "a sample of blood once a month for three months." In truth, blood was being drawn from the test subjects far more frequently.

Benda's letter cast the research project in an innocuous, if not a valuable and festive, light. Study participants would receive such privileges as a "quart of milk daily," trips to baseball games, the beach, and outside dinners. In short, according to the letter's author, the participants "enjoy it greatly." Benda, in fact, actually told parents to rearrange their vacation plans so that test subjects would not leave the institution during a critical stage of the research.[42]

Fernald medical personnel maintained files on each research subject, including whether parental permission for their child's participation had been received. Some documents identified who had granted permission, and others disclosed personal information about the test subjects.[43] In a number of cases "no relatives" was handwritten after a subject's name, meaning he was declared a ward of the state, a designation that gave the superintendent total control over a boy's fate.

As Benda became more confident of his authority and more experienced in his position as Fernald's medical director, he expanded his own scope of research and embarked on some projects that would give considerable pause today.[44] One of the more ghoulish was Benda's desire in September 1953 to use a very sick child for a calcium metabolism experiment that would require giving "a dose of 50 uc Ca 45 to a moribund gargoyle patient." (The terms "gargoyle" and "gargoylism" were commonly used in the medical community during the first half of the twentieth century to describe Hurler-Hunter syndrome, a degenerative disease of the nervous system associated with dwarfism, extreme disfigurement, and death by the age of ten.) In his request letter to the AEC, Benda claimed that the ten-year-old boy had "suffer[ed] from this severe metabolic disorder since birth but [was] going progressively downhill at present." The patient was considered to have only a "few months" to live, and Benda urged "prompt consideration" of his request so his research could begin without further delay.[45]

He bolstered his argument by reminding the AEC of his three-and-a-half-year experience of conducting calcium metabolism studies with MIT researchers, and noting that permission had previously been granted other medical investigators "for the use of higher doses administered to moribund patients."[46] The study was never completed due to the death of the patient on day 16 of the study. Interestingly, despite injecting the boy with a significant amount of radioactive material, the authors of the journal article detailing the study expressed their "appreciation to the staff of the Walter E. Fernald State School who nursed this patient with skill, self-sacrifice and patience."[47] Though one cannot be certain, it is unlikely that Benda or anyone else at Fernald informed the young boy's parents what they had in mind for their terminally ill son.

The level of knowledge government officials had regarding medical research projects taking place at Fernald as well as throughout the rest of the commonwealth's mental health system is unknown, but there are some indications that they periodically sought out such information. For example, in 1956 Dr. Jack R. Ewalt, commissioner of the Massachusetts Department of Mental Health, informed institutional superintendents throughout the state that "the Governor and some members of the legislature expressed interest in the kinds of research going on in our institutions." Eighteen months had passed since the last information had been provided, and Ewalt was requesting another accounting.[48]

Outside of the handful of former State Boys who are around to tell their story, few witnesses or researchers are alive from the early years of the Fernald experiments—clinical trials that would eventually stretch over three decades. One of the very few who was present at the program's inception and someone who has remained steadfast in his conviction that the experiments were unassailable scientifically and safe from a health perspective is Dr. Constantine Maletskos. Widely sought out by the media at the time of the Fernald revelations in 1993, Maletskos assured everyone, "I feel just as good about it today as the day I did it. The attitude of the scientists was we're going to do this in the best way possible," and those chosen as test subjects would receive "the minimum radiation they could possibly get and have the experiment work."[49]

Little has occurred since those remarks to alter his view that MIT's research at Fernald was ethically sound, of scientific merit, and done with the utmost care.

In our series of interviews with Maletskos in 2011, he underscored his chief concern at the time—that the test subjects receive "the littlest dose of radiation possible" to ensure that no one was harmed, but still sufficiently large to accomplish the goals of the study. The issue—to "better understand young kids and digestion"—was the crux of the study.[50]

Maletskos was a graduate student doing research in MIT's biology department when Robley Evans selected him to help coordinate the Fernald study. Maletskos felt honored to be chosen to work on a project of significance. It was an exciting time at MIT. A vast array of scientific investigations was under way, and human experimentation was part of it. On the advice of Karl Compton, another giant in the nascent field of nuclear physics and the president of MIT, Evans was recruited after the war to head up MIT's nuclear physics department and oversee the construction of one of the nation's first cyclotrons. There was an "explosion of interest," says Maletskos, in nuclear issues, its implications for science, and how it would affect mankind in the postwar years.

And test subjects were not misused. "We treated the kids great," said Maletskos, recalling his interactions with the children. He would go on to spend many years in scientific research, train doctors in the use of radioactive material, and eventually do similar isotope studies on the elderly. Maletskos remembers Clemens Benda as more than competent; "he was a smart cookie," somebody who knew his business. "We wanted to know how people metabolize certain minerals like iron and calcium. We needed a controlled population for the study," said Maletskos, describing a group that could be confined, monitored, and fed easily, and one in which urine, feces, and blood could be regularly collected.

"Everybody was trying to do the job right," recalls Maletskos of the mind-set that pervaded the study. "There were no rules in those days. We made the rules as we went along. I had no knowledge of the Nuremberg Code. The Code had little to no impact. I don't recall it being taught in school or mentioned in classes. I had to set my own standards and believed the littlest you can use [radioactive isotopes] to gain the information you want, the better. I did it just that way; I was measuring very low amounts of radiation. But it worked and we got some nifty numbers."

After the passage of many decades, however, Maletskos admits yesteryear's standards do not measure up to contemporary rules and expectations regarding clinical research. "I can see in retrospect it was not right," he now says.

The research from Maletskos's early nutrition studies at Fernald would eventually be published in four scientific journal articles in the early and mid-1950s.[51] If anyone reading the articles at the time thought there was something objectionable about using institutionalized, mentally challenged children as test subjects in experiments incorporating radioactive material, there is no record of it. As Maletskos says, it was an era that downplayed or was oblivious to rules, regulations, and ethical constraints; researchers had wide latitude in crafting and orchestrating their research protocols.

More than a decade would pass before Henry Beecher published his groundbreaking piece in the *New England Journal of Medicine* illuminating colleagues' ethical lapses in the conduct of their research.[52] During much of that time, the Fernald Training School and Wrentham, its nearby sister institution, would continue to host medical experiments using their wards as free and compliant research material.

Throughout the 1950s and 1960s, thyroid studies became the focus of researchers' interest at the two institutions. In order to better understand how the thyroid gland functioned, a series of experiments was initiated incorporating radioactive iodine as a tracer to track and monitor the passage of iodine in the body. Interestingly, the studies utilized not only Fernald residents but their parents and residents of Wrentham. Studies that focused on one specific birth defect were not unusual. In the mid-1950s, for example, researchers from Harvard Medical School and Beth Israel Hospital traveled to Fernald to administer "70 microcuries of . . . radioactive iodine" to "21 mongoloid subjects" ranging in age from five to twenty-six.[53]

However, the Massachusetts Commission investigating the Fernald experiments discovered that "the amount and type of tracer materials used went beyond the minimal tracer levels of the earlier nutritional research studies" and demanded immediate follow-up examinations of the former test subjects.[54]

The Massachusetts Investigative Task Force found one study particularly troubling. Apparently the 1961 study of children at Wrentham was designed to determine the amount of normal iodine that had to be added to children's diets to block the uptake of radioactive iodine they might be exposed to from nuclear fallout after a nuclear attack or accident. This study had definite military overtones and was subsequently referred to as a "Cold War experiment." The task force was also to learn that "the results of this study were actually referenced at a 1989 European conference after the nuclear accident at Chernobyl."[55]

Seventy young children at Wrentham participated in this study, almost all between one and eleven years of age. According to task force findings, over sixty of the children were given daily dietary supplements of sodium iodide, and others received separate doses every two weeks for three months. In total, each of these subjects received eight microcuries of radioactive iodine. The task force members assumed that other subjects in the study had received similar doses. According to the task force, the study's designers "chose this population of children because it was desirable to secure children living under constant conditions of environment, diet, and iodide uptake."[56] The Wrentham study was coordinated by researchers from Harvard Medical School, Massachusetts General Hospital, and the Boston University School of Medicine, and it was supported by the Radiological Health Division of the US Public Health Service.

Subsequent studies at Wrentham during the mid-1960s included dozens of children between the ages of one and fifteen who suffered from Down syndrome and other forms of mental retardation. One study was designed to test the thyroid function in children with Down syndrome and compare it with that in normal children.[57] Additional studies were undertaken at Fernald and Wrentham that measured thyroid function in both myotonia dystrophica, a severe chronic neurological disorder, and Down syndrome. Benda, who spent years exploring the cause of "mongolism" and its relationship to thyroid dysfunction, was the lead author of the myotonia dystrophica study along with Constantine Maletskos.[58]

The task force collected the opinions of several experts regarding the short- and long-term health effects of exposure to radioactive tracers. Calling them "minute amounts (less than one billionth of an ounce) of radioactive iron and calcium" and often comparing radiation exposure levels to those of people living in different geographic locations, flying in airplanes, and having diagnostic medical and dental X-rays, the experts agreed that the likelihood of any of the test participants developing cancer was very small, though the greatest risk of leukemia resided with those young children who were part of the thyroid gland studies.[59] The experts did, however, have serious reservations about the "informed consent" aspect of the studies and agreed that the experiments would "not be permitted today even with the low radiation doses that were used and informed consent of parents or guardian."[60]

Regardless of the reassurances from the medical experts concerning long-term health risks, many of the former State Boys who were part of the Science Club remain distrusting of the medical community, and convinced that much more went on at Fernald than was disclosed. "The report said we weren't hurt by the study," said Joseph Almeida, who spent his youth in Fernald and is outspoken in his denunciation of its treatment of children. "They worked for the state. What do you expect them to say?"[61]

"There were people buried out there in paupers' graves beyond the farm," said Gordon Shattuck, who continues to decry the institution's cavalier use of children in assorted medical experiments over the years. "They just put 'em in the ground in a pine box. They killed them. Benda was working on them all the time. They died from the medical experiments."

As Austin LaRocque told us, "Fernald ruined a lot of kids' lives. They were the worst years of my life. They stole my childhood."

"We were told we were morons and idiots all the time," said Charlie Dyer. "Nobody wanted us or cared about us. They just used us for whatever they wanted or needed done."

The state investigative task force issued its final report in the spring of 1994, but it received decidedly less media coverage and public interest than the original allegations of abuse at Fernald.

Surprisingly, the report omits any questions regarding the scope of the investigation—the commission only examined medical research incorporating radioactive isotopes. Was no one curious about the possibility of other types of medical research occurring at Fernald over the years? During the era of unrestrained research, institutions housing vulnerable populations were a treasure trove of investigative opportunity for medical researchers and pharmaceutical companies. One trial invariably led to a second, and then a third, and fairly soon a cottage industry of medical research was in full bloom with too many players benefiting for the practice to be curtailed. For Fernald to have escaped this common pattern would be unusual.

Sandra "Sunny" Marlow, the librarian at the Howe Library at Fernald who is credited with discovering the papers disclosing the school's long association with medical researchers and the radioactive tracer studies, said she was not impressed by the commission's data collection effort. In short, "They were not particularly aggressive in collecting information."[62]

More damning yet is the account of Dr. Doe West, a former chaplain at Fernald and the coordinator of the Massachusetts Task Force report, who expressed reservations at the time and regrets the parameters placed on the investigation. West believes she had discovered evidence of "earlier studies at Fernald," more than rudimentary diet research incorporating radioactive isotopes. "There was clearly more going on than Cream of Wheat versus Quaker Oats," says West.[63] She was informed in no uncertain terms, however, that nutrition studies and the use of radioactive material as tracers would be the focus of the investigation. Anything other than that was to be jettisoned. Curiously, no one, neither lawmakers, citizens, nor members of the task force—or even patients' relatives—protested the narrow scope of the investigation.

Nearly two decades have passed since West performed the many daily duties of coordinating a state-authorized investigative commission, but she still expresses regret that a more comprehensive study was not undertaken. "There was massive proof that Fernald was ground zero for a wide spectrum of maladies," says West. "The researcher/doctor role was blurred and doctors used the school as a training ground" for students and research projects. "Doctors wanted to come there to do their research. It was convenient," and it also had "affiliations with top schools, and parents who were desperate or disinterested. They didn't challenge authority."

By deceiving parents with incomplete and misleading claims, pitching seductive offers to disadvantaged and affection-starved children, and punishing those who chose not to participate, physicians circumvented moral principles for the sake of convenience and their conception of the greater good. Eugenics and the Cold War had contributed greatly to developing a flexible moral landscape that rewarded utilitarian action over ethical restraint. Though the Fernald story is one of the more poignant and better-documented cases from the Cold War era, it actually pales in comparison to many other experiments that placed institutionalized children at even greater risk.

**REGARDLESS OF WHETHER CLINICAL TRIALS** during the Cold War were designed to learn more about radiation or the nature of disease, institutions holding impaired and orphaned children often became the epicenter for the investigational studies. The Manhattan Project and the extraordinary campaign to build the world's first atomic bomb during World War II touched off a frenetic effort to learn the secrets of a new science and weapon that had never existed before. Decades earlier, scientists had known that radiation was dangerous; close proximity to X-rays by early investigators had confirmed their threat, and most knew that women who ingested liquid radium while painting watch dials were subject to early and painful deaths. Now, during the last years of the war, Manhattan Project doctors raised the alarm about potential problems. Enriched uranium and plutonium were necessary for the project's success, but they were both problematic from a health standpoint. As one scientist counseled his colleagues, plutonium even at low doses should be regarded as "extremely poisonous." Additional research was needed, and that required human experimentation.

For example, in medical journal articles assessing thyroid activity, the subjects of the research were often described as those "institutionalized for mental inadequacy," "mental defectives," "juvenile delinquents," and "abnormal children." For example, one study measuring thyroid activity in children using radioactive iodine (I–131) at the dawn of the Cold War in 1949 incorporated both "normal" and "abnormal" children. Not surprisingly, the abnormal group formed the majority of the subjects and consisted of a cross-section of children including a "cretin," a "pituitary dwarf," and others who were said to possess characteristics of "mongolism" and "gargoylism," terms that eugenicists decades earlier had helped to popularize.[64] For example, a month-old infant described as a "mongolian idiot" was presumed to be one of the first with that malady to be injected with radioactive iodine. The research was conducted at the University of Michigan Medical School and funded in part by the American Cancer Society.

Radioactive studies were conducted on even younger subjects. At the University of Tennessee College of Medicine in 1954, Dr. Van Middlesworth experimented on seven newborn boys—"six Negroes" and "one Caucasian"— just two and three days old and weighing between seven and nine pounds.[65] Middlesworth believed that no such studies had been reported on newborns. Realizing that "the use of radiation in the very young organism is open to some question," he decided to consult a local group of advisors. The group—a radiologist, a radiation physicist, two internists, two pediatricians, a physiologist, and a pathologist—decided that injecting I–131 into newborns was acceptable and "not expected to be harmful." Middlesworth stated that he described the procedure to the mothers of the infants and received their consent. Whether mothers—African American or white—in the South in the early 1950s understood anything about radiation uptake studies and the half-life of certain atomic particles is another issue. Doctors were hopeful, however, that the exercise would "prove useful in the diagnosis of thyroid abnormalities in infants" in the future.[66]

Just a year later, doctors at a Michigan hospital increased the number of infant test subjects receiving radioactive iodine to sixty-five. Most of the infants were less than two weeks old. "The possible danger of using I–131 on a premature infant "was debated among the director of research, the chief of the Radiology Department, the director of the Radioisotope Laboratory, the director of the Pathology Department and the chief of the Pediatric Division. The consensus was that oral administration would not be harmful.[67]

The accumulation of knowledge was a powerful incentive and the availability of radioactive isotopes to foster that goal spurred excitement. Thyroid gland investigations in particular proved attractive and were much replicated, despite potential health risks from radiation exposure. As doctors in one journal article soberly stated, "A calculated risk is taken in performance of any irradiation examination, but the potential benefits may outweigh any hypothetical dangers whenever such an examination is indicated."[68] The question naturally arises: the potential benefit for whom? The subjects in this study were normal newborns from 72 to 180 hours old. Radiation exposure at this young age could only do them harm.

Even into the 1960s, researchers were still traveling to custodial institutions to access test subjects. The attraction was not a secret; doctors freely admitted it in subsequent journal articles. As one doctor frankly wrote, "We chose this population of children living under constant conditions of environment, diet, and iodide uptake."[69] The investigators who were interested in the impact of atmospheric radioactivity from nuclear explosions and desirous of conducting a study on the suppression of the thyroid function to prevent radioactive iodine retention could have just as easily conducted their experiments in their own backyard. There was no shortage of schools with the controlled environments the researchers sought. But Wrentham and Fernald and others like them had something else: throwaway children who had been given up by their families and for all intents and purposes severed from society. If something untoward should happen during a study, there was far less chance of hearing from an outraged parent, dealing with an embarrassing news story, or encountering any uncomfortable legal entanglements.

Gordon Shattuck, Charlie Dyer, and Austin LaRocque eventually learned of their history as human guinea pigs during the Cold War and the Golden Age of Medical Research. The vast majority of such subjects, however, would never discover how they had been used and conscripted as "volunteers" in the campaign to benefit medical science.

Charles Davenport was the founder of the Eugenics Record Office and an indefatigable champion of the better breeding movement. Consumed by the goal of ridding society of "defectives," Davenport regularly journeyed to Letchworth Village to study abnormal inherited traits. He would initiate the castration of a "mongoloid dwarf" to augment that research. Photo courtesy of the National Library of Medicine.

Over-crowded and under-staffed state institutions for the cognitively impaired and developmentally disabled, such as this one in New Lisbon, NJ, were often inviting research sites for those physicians desirous of performing a wide range of clinical trials. Photo used with the permission of Philadelphia Inquirer Copyright 2013. All rights reserved.

One year old Mark Dal Molin with his sisters: Chris (3), Karen (4), and Gale (2). Born with cerebral palsy, Mark was placed in Sonoma State Hospital in 1958. He would die there in 1961, according to his sister's investigation of the case, after being used as a test subject in radiation studies. Photo courtesy of Karen Dal Molin.

Dr. Jonas Salk, the celebrated University of Pittsburgh virologist who developed the first effective polio vaccine, coordinated some of the vaccine's early trials at the D. T. Watson School for Handicapped Children and the Polk State School. Photo used with the permission of Philadelphia Inquirer Copyright 2013. All rights reserved.

Dr. Hilary Koprowski tested his live–virus vaccine on children at Letchworth Village in 1950. Koprowski's actions shocked many of his colleagues and his use of the term "volunteer" to describe young test subjects drew critical comment from the The Lancet. *Photo used with the permission of Philadelphia Inquirer Copyright 2013. All rights reserved.*

Saul Krugman was an accomplished NYU virologist whose research on hepatitis from the 1950s to the 1970s would break new ground in understanding and fighting the disease. His research on retarded children at Willowbrook would also become some of the most controversial experiments of the last half-century. *Photo courtesy of the National Library of Medicine.*

When Albert M. Kligman earned his MD, he moved from plant fungus to human fungus. Much of his clinical research in the late 1940s and early 1950s incorporated children at the Vineland and Woodbine Institutions for the Feebleminded in southern New Jersey. *Photo courtesy of Allen M. Hornblum.*

Roman Catholic
confirmation of Austin
LaRocque and his sister
Rosie at the Fernald State
School in the early 1950s.
Archbishop Cushing of
Boston presided over the
service, a rare event for the
children who were provided
little education but made to
perform an array of menial
tasks. Gordon Shattuck
is at the extreme right.
Photo courtesy of Austin
LaRocque.

Dr. Clemens Benda emigrated from
Germany during Hitler's rise to power in
the 1930s. He would become the Medical
Director of the Fernald State School from
the late 1940s to the mid-1960s. During
that time he would help MIT establish the
"Science Club" at Fernald and pursue the
causes of "mongolism." Photo courtesy of the
private archives of Rev. Dr. Doe West, from
her work and studies in disability rights.

*Shy, withdrawn, and the son of an institutionalized parent, six-year old Ted Chabasinski was placed in Bellevue Hospital, declared schizophrenic, and subjected to a battery of electro-shock treatments.*

*Lauretta Bender was one of America's leading neuropsychiatrists during the middle decades of the twentieth century. In her quest to combat childhood schizophrenia, she would become an ardent advocate of putting institutionalized children on electroshock and LSD regimens. Photo courtesy of the National Library of Medicine.*

Walter Freeman, a George Washington neurologist, would become the nation's foremost lobotomist. A disciple of Portuguese physician Egas Moniz, Freeman never lost an opportunity to proselytize the attributes of lobotomy in the fight against a wide range of psychological maladies. He once performed lobotomies on 25 women in one day. Photo courtesy of the National Library of Medicine.

Former "State Boys" Charlie Dyer, Austin LaRocque, and Gordon Shattuck reunite on the grounds of the Fernald School. As children six decades earlier, they were inmates of the institution, forced to work for their keep, and used as unwitting test subjects in medical experiments. Photo courtesy of Allen M. Hornblum.

# EIGHT

# PSYCHOLOGICAL
# TREATMENT

### *"Lobotomy . . . Is Often the Starting Point in Effective Treatment"*

**SEATED ALONE AND TERRIFIED IN A COLD AND GLOOMY HOS-**pital corridor, six-year-old Teddy Chabasinski tried to remain calm, but everything about New York City's Bellevue Hospital frightened him. As opposed to the other boys his age housed on one of the hospital's wards, he was relegated to a bed in a bleak, lonely hallway. The dreary ceilings seemed as high as the sky, and the huge windows were covered in urban dirt and grit. The smelly, filthy mattress he was forced to sleep on disgusted him, and an old, heavily stained olive drab blanket left him nauseous and shivering throughout the night.

It was winter 1944, and young Ted's life had taken a terrible turn. A bright boy who loved to read and listen to the news on the radio, Ted knew about the surprise Japanese attack on Pearl Harbor, victory gardens, and the war against the Axis powers. But as he sat alone and shivering in a hospital corridor, he would now be involved in his own life-and-death struggle, a struggle so severe that on occasion he wished he were dead.

Ted had been taken to Bellevue after being snatched from his foster parents by Miss Callaghan, an officious social worker who thought he displayed all the characteristics of someone with a serious mental illness. He was smart and had learned to read much earlier than other children his age, but Ted always seemed withdrawn, he cried easily, and he ran away from older children. He also fought with his younger sister. Ted's mother was schizophrenic, unmarried, and institutionalized, which resulted in his being sent to a foundling hospital when he was just ten days old. Five months later he was handed over to foster parents. "They were nice but not strong," Ted says of his foster parents. "They didn't know how to stand up to the social worker." Miss Callaghan, according to Ted, figured that if Ted's mother was schizophrenic, he must be, too.[1]

Ted was deposited in the children's ward of the psychiatric division of Bellevue Hospital in Manhattan. He awaited his fate fretfully as children his age were probed, prodded, and examined for the cause of their psychological disturbance. Ted never understood why he was there or why no one came to his rescue at night when he screamed for help while being sexually assaulted by a hospital attendant. No one seemed to care, not even the female doctor of whom everyone appeared to be in awe. Dr. Lauretta Bender occasionally came down the hallway with her entourage of aides and acolytes. "They seemed to worship her," Ted would write years later, though he believed some might, like him, actually have been in fear of her. "Sometimes she would pass very close to me," he would write, "but never acknowledge me. It was as if I didn't exist. She seemed to just look right through you."[2]

Ted remembers being taken to a room down the hall where attendants forced him to lie on a gurney. A rag was shoved in his mouth and down his throat till he nearly choked. It was then that he saw Dr. Bender, the person who had formerly diagnosed him as suffering from childhood schizophrenia. He recalls little else of that first electroconvulsive therapy (ECT) session.

He'd wake up in a dark room, dizzy, disoriented, with a terrible headache and little memory of what had happened. Sometimes he wouldn't even remember his name. Sometimes another boy, Stanley, would be in the room. Stanley was a large boy about thirteen years of age who never talked or moved. Ted became terrified of him though he didn't really know why. He referred to his lost memory as a "black hole that the shock had created." To combat the memory loss, he started to focus on his name each time he was dragged into the ECT room. "I'm Teddy, I'm Teddy, I'm here in this room, in this hospital," he would tell himself over and over as he was being strapped down. The pain, the disorientation, and the seemingly endless sessions of being shocked and then trying to remember who he was and why he was being punished this way would go on until May of that year.

During that long, harsh winter, Ted learned the system pretty well. On those mornings he was not given breakfast, he would begin to cry; he knew he could expect another shock treatment. He would then be dragged screaming into the room where Dr. Bender or one of the other doctors would shock him with electricity. Though he resisted with all his might, he was no match for the three or four attendants who were always there to restrain and secure him with leather straps. Then bolts of electricity were shot through his brain until he was unconscious. ECT patients—normally adults—were strapped down, a rubber object was forced between their teeth so they wouldn't bite the tips of their tongue off, and electric coils were attached to the patient's temples. When the switch was thrown and the current shot through the patient's brain, a series of violent convulsions would begin. It was not uncommon for patients to leave the gurney with a broken bone or dislocated joint along with a tremendous headache and loss of memory.

Ted often wished he were dead. Ironically, he and the other boys were frequently "marched across the hall to the girls' ward where they were supposed to sing and show how happy and normal they were." Ted rarely complied, a sign, he was told by nurses and attendants, of his illness. If he wanted to get better, he had to sing. More often than not, tears would streak his face as he did so.

Though he was only six years old, intelligent, and never violent or disruptive, Ted was given the usual complement of twenty ECT treatments. He would never understand what crime he had committed, what he had done wrong, or why he had to receive such a horrible and oft-repeated punishment. As he would write many years later, "And so, in May 1944, after being raped and killed over and over, I finally was released from Bellevue. The little boy who had been taken there to be tortured didn't exist any more. All that was left of him was a few scraps of memory and a broken spirit, and the rest was ashes in a giant dark pit, mixed with the ashes of the hundreds of other children who had been tortured and burnt alive by Doctor Bender, a leader of her profession."

Ted was returned to his foster family in the Bronx, but he was not the same boy whom the authorities had taken away just a few months earlier. Terrified, he rarely left his mother's side. For a while he refused to go outside. When reintroduced to his tricycle, which earlier he had confidently ridden through the neighborhood, he was apprehensive. Where he once knew every block and every house on the street, he was now dumbfounded. "Suddenly," he would later write, "I realized that I didn't know where I was and I panicked. I used to have a sense of freedom, that I was a big boy and could ride my bike anywhere, but that was gone now."

Geography and familiar physical landmarks were not the only things the shock treatments had stripped from Ted's brain; people had also been cut out of his memory bank. Karl, one of his closest friends who lived just two houses away, was now totally unfamiliar to him. "I didn't know who he was. Miss Callaghan said this was a very bad sign. It meant I wasn't getting better. She told my foster parents my memory loss was due to my malady, not the shock treatments."

Once again Ted was taken from his foster parents. This time he was placed in Rockland State Hospital in Orangeburg, New York, where there were other children—some who were forced to wear football helmets because they banged their heads against the wall—and adults who he realized were "crazy." Many seemed dazed and delusional while others required restraints. Ted would no longer be confronted with ECT, but he was still a prisoner and fearful of his new surroundings and the people there.

Classified a schizophrenic at seven years of age, Ted would remain at Rockland State for ten years until he was seventeen and deemed cured. During that time he would observe all manner of personal degradation and institutional abuse. Incredibly, he survived his ordeals. Adjusting to the free world after many years in a madhouse was not easy, but he managed to make a life for himself.

Though he never earned a high school diploma, he enjoyed reading and learning, and subsequently matriculated at New York University, where he accumulated thirty-eight credits before having to drop out due to lack of funds. City College of New York presented an attractive economic alternative, so he enrolled there and eventually earned both a GED and a bachelor's degree in psychology in 1961. His excellent grades also earned him a Phi Beta Kappa key. He then went on to take graduate courses at Columbia and New York

University. Years later he would move to the West Coast, enter and graduate from law school, and help lead a campaign in Berkeley to stop the use of ECT. The municipal ballot measure passed by a 62 percent to 38 percent vote.

Over a half century after being diagnosed with childhood schizophrenia and having to endure nearly two dozen shock treatments, Ted Chabasinski is as adamant as ever about the destructive scientific impact of doctors caught in the wake of novel cures and professional ambition. Not surprisingly, he is especially critical of Dr. Lauretta Bender. Her attachment to the latest but unproven medical techniques and potions had earned her tremendous recognition and honors, but no doubt at great cost to those who became her patients and experimental subjects. "Doctors were an elite group," says Chabasinski, "and as one of the rising stars in the field, Lauretta Bender was shocking subhumans, mental patients, people who were not of the same value as real or regular humans."

Though he knows she is still respected in many medical and psychological circles, Chabasinski roundly criticizes Bender for her inhumanity, elitism, and attraction to untested therapies that often did more harm than good. Devalued, institutionalized children and adults were grist for the exploratory mill Bender established, and professional colleagues who should have known better praised her or remained silent about her misguided ventures.

"Bender is comparable to the Nazi doctors," argues Chabasinski. "I was helpless, and nobody in society cared about me. I was a shy, somewhat withdrawn kid, but I wasn't crazy or schizophrenic or in need of institutionalization. But Bender needed ECT subjects. They were able to do what they wanted with no thought or care of the consequences." Chabasinski blames much of the laissez-faire atmosphere in the medical arena on the nation's fascination with the eugenics movement earlier in the century that allowed Bender and people like her to exploit the "disenfranchised." "The attitude that some people were subhuman took hold," says Chabasinski, "and a campaign began to get rid of them. And if people are declared subhuman, you open the door to all sorts of stuff. It's not a very large step from there to human experimentation."

**BORN IN BUTTE, MONTANA**, in 1897, Lauretta Bender faced her own struggles as a child. The fact that she repeated first grade three times led some to believe "she was mentally retarded." Her habit of reversing letters when reading and writing was due to dyslexia, it turned out, a handicap she over-came along the way to becoming her high school's valedictorian.[3] She would go on to earn degrees at the University of Chicago and Iowa State University and specialize in psychiatry. Bender quickly came to believe that "learning disorders were determined neurobiologically and were related to delayed maturation of brain function that were required for language."[4] Though she may be best known for her role in the creation of the Bender-Gestalt Test used to examine personality and emotional problems, she also gained acclaim for work and theories regarding language development and reading disability. Her much-used diagnosis of childhood schizophrenia gained her increased fame—that and her fervent attraction to questionable remedial devices and potions to harness those maladies is what many find troubling today.[5]

Bender joined the staff of Bellevue Hospital in New York City in 1930. She was the senior psychiatrist in charge of the Children's Services there for two decades before moving on to other influential positions. During those early years of the Great Depression and into the 1940s at Bellevue, Bender was associated with innovative hospital programs like puppet shows and music therapy. Her ardent championing of ECT and her routine and cavalier over-use of it with children cries out for critical comment.

As Bender repeatedly wrote in journal articles, "Childhood schizophrenia is an early manifestation of schizophrenia as it appears in adolescents and adults," emphasizing its "inherited predisposition," "noxious or traumatic events," and their impact on "maturational lags or embryonic immaturities."[6] Unfortunately, particularly given her growing status and influence in the medical and psychiatric community, she tended to see schizophrenic characteristics in an overwhelming number of the children she examined. Ted Chabasinski was one of them. Chabasinski might have been timid, insecure, and in need of counseling, but it is highly unlikely that psychiatrists today would have labeled him schizophrenic and placed him in an institution. Nor would various forms of shock treatment be prescribed for him today.

Bender was an early and zealous devotee of all manner of convulsive treatments. The 1930s, a period that coincides with her early years working with children at Bellevue, would be a fruitful period for the creation and use of such treatments. She and other doctors struggling to diagnose and treat a wide range of mental problems with a small armamentarium of cures eagerly welcomed whatever new therapies came along. Dealing with a perpetual flow of profoundly disturbed people with only minimal understanding of the mind, its biology, and its inner workings could not have been an easy undertaking. According to medical historian Joel Braslow, "The shock therapies provided the first new remedy since the introduction of malaria fever therapy in the 1920s." Hailed by the press as "wonder cures of modern medicine," these new practices acted directly on the body.[7] Doctors became enamored of them; that was particularly true of Lauretta Bender.

Manfred Sakel, a Viennese physician, invented insulin shock therapy in 1933 when he realized that massive doses of insulin would precipitate hypoglycemic shock. Sakel discovered that when large doses of insulin were given to psychiatric patients, some showed dramatic improvement. Though he would never completely understand why a low blood sugar shock would return some psychotic patients to normalcy, Sakel's new technique quickly drew converts, despite occasional failures in which patients died on the table of heart failure and cerebral hemorrhage.

Metrazol convulsive therapy came along a year later and proved an even more powerful but hazardous form of shock therapy. Patients had a deathly fear of it and pleaded with doctors to be taken off Metrazol regimens. Pentylenetetrazol, the generic name for the drug, was discovered by Ladislas J. Meduna in Budapest. He realized that intramuscular injections of camphor-related drugs caused intense convulsions. Quicker acting, cheaper, and requiring less manpower than insulin, Metrazol therapy soon rivaled insulin therapy at many hospitals. There would be a tidal wave of complaints, however.

In 1936, two Italian physicians came up with another method to induce seizures, this one using electricity. Ugo Cerletti and Lucio Bini had been electrifying animals—and killing a good number of them—before they perfected their machine and learned where the electrodes should be placed. By the end of the decade, psychiatrists had three different shock techniques to treat melancholic patients, depressives, and psychotics. But they also used them on a great many others whose symptoms and maladies made them highly unlikely candidates for convulsive therapy.

Doctors and hospitals, overrun with mental patients, quickly bought into one or the other or all three methods of shock therapy. In 1941, according to Braslow, "42% of institutions surveyed had electroshock machines just three years after the first human electroshock trial."[8] By the end of the 1940s, ECT would be the backbone of therapeutic care at asylums and mental hospitals across America. At Stockton State Hospital in California in 1949, for example, "patients receiving electroshock increased five fold over the previous year to 7,997. Doctors were shocking over 60% of the patients at Stockton."[9]

Lauretta Bender was one of the earliest and most enthusiastic proponents of the new shock-therapy regimens, especially when it came to prescribing such treatment for children and publishing articles on her studies. Assigning young children to "convulsive therapy," however, could easily strike some objective observers as torture rather than treatment.

Granted, insulin and electric shock were less physically jarring than Metrazol and resulted in patients having less need of orthopedic attention, but each method had its drawbacks, and none was guaranteed to produce a cure. Yet Bender clearly believed that such measures were not only safe but also extremely beneficial to patients. As she wrote in one article, the "treatment does give a spurt to lagging maturation, it stabilizes the functions of the autonomous nervous system and the electroencephalogram." In addition, shock therapy broke up "bizarre schizophrenic patterns in body image . . . improved level of intelligence, gestalt drawing," and "personality improvement." In short, Bender believed, "The prepuberty child is more normal" with the use of shock therapy.[10]

Bender, in fact, became so enamored of convulsive treatment that she argued that ECT was appropriate for young autistic children, a treatment that would strike today's therapists and psychiatrists as bizarre. More confounding yet, especially from the perspective of young subjects like Ted Chabasinski who experienced Bender firsthand, is her statement that "a close personal relationship with a physician during the period of recovery from the insulin, while the child is fed sweets, may encourage a child to talk and it may be generally therapeutic."[11] Chabasinski's recollections of Bender revolve around her "cold, imposing stare," her indifference to him as she quickly marched through the corridor during hospital rounds, and her "throwing the switch" that caused him to convulse and lose consciousness. He has no memory of being given candies, watching puppet shows, hearing a kind, reassuring word, or receiving a cold compress for his throbbing, sweaty brow.

Bender's fondness for shock therapy in all its forms—"it stimulates appetite and general well being"—was not shared by the children Chabasinski knew at Bellevue and Rockland State Hospital. It was Dr. Bender, however, who was writing the journal articles and trumpeting the technique's perceived success stories. In one 1947 article titled "One Hundred Cases of Childhood Schizophrenia Treated with Electric Shock," Bender discussed the results of giving "grand mal seizures" to children whose IQs ranged from 44 to 146. All were diagnosed schizophrenic and supposedly suffered no lasting negative effect from the treatment on subsequent psychometric tests. As she unscientifically proclaimed about ECT therapy, "the children were always somewhat improved by the treatment inasmuch as they were less disturbed . . . more mature . . . happier."[12] Chabasinski would certainly disagree with that assessment and so would a good portion of today's pediatric community.

From its earliest days, shock therapy was fraught with controversy, and opposing camps argued the new remedy's strengths and weaknesses. Sound, unbiased, and controlled studies of shock therapy were few and far between. Anecdotal tales of dramatic recovery seemed to rule the day. Such stories only encouraged greater use of the new remedy. Just prior to World War II, a US Public Health Service survey disclosed that nearly three-quarters of 305 public and private institutions were using insulin therapy.[13] Lauretta Bender needed no encouragement; she was a devoted advocate of shock therapy from the very beginning and went further than most in prescribing it for young children she considered schizophrenic.

At the time there was little direct criticism of Bender's penchant for sending electricity through the brains of children, but opposition would soon grow, and Bender would come under direct attack. In a report for psychoanalysts organized by Karl Menninger, Bellevue Hospital's ECT program for preadolescent children with schizophrenia was called "promiscuous and indiscriminate."[14] The report argued that shock therapy showed some effectiveness with depression but not with other mental diseases such as manic depression or schizophrenia. More important, the report said that brain damage was an inevitable consequence of ECT: "Abuses in the use of electro-shock therapy are sufficiently widespread and dangerous to justify consideration of a campaign of professional education in the elimination of this technique, and perhaps even to justify institution of certain measures of control."[15] The American psychoanalytic community had spoken, but it would back off in coming years as those supporting a physical/biological school of thought mounted considerable counterpressure.

Bender did not escape the sting of criticism. Her decision to subject children as young as four years old to a battery of twenty ECT sessions was questioned as was her unsubstantiated belief that children "had become much more sociable, composed, and able to integrate in group therapy as a result of the daily ECT."[16]

It is difficult to say with any certainty how Ted Chabasinski and the many children like him with relatively minor behavioral issues who were labeled schizophrenic, institutionalized, and given ECT would have done if they had escaped the clutches of doctors and institutions with their own agendas. Some might have outgrown their problems as they approached adolescence. Others might have required counseling and medication. But shock treatment for young children should have been a last resort, if an alternative at all.

**INSULIN, METRAZOL, AND ELECTRIC SHOCK** therapies were not the only new remedies for damaged minds to emerge out of Europe in the 1930s.

Egas Moniz was a Portuguese physician, and like many of his medical colleagues, he decried the slim armamentarium of effective weapons in the fight against mental illness. For years they had been treating patients suffering a range of phobias, delusions, weeping fits, lack of self-restraint, and various forms of depression. Cures were hit or miss and usually only temporarily effective. Moniz believed something more radical was needed to address the problem, something that would dramatically shake up the brain cells that fostered bizarre behavior. After returning from the International Neurological Congress in London in 1935, he decided to attack the problem directly. He created a scalpel-like metal instrument in the shape of an apple corer and instructed his colleague, surgeon Almeida Lima, to cut into the brains of nearly two dozen anxiety-ridden and insane patients.

The results, according to Moniz, were nothing short of miraculous. Hypochondriacs were no longer preoccupied with contracting cancer and polio, the depressed jettisoned notions of suicide, and those in fear of imaginary pursuers no longer suffered from persecution complexes. Granted, most of the long-documented cases of dementia praecox—an early term for schizophrenia—seemed intractable, but there were enough significant changes in other patients to capture the attention of the medical community. Perhaps, some were willing to concede, Moniz was on to something; perhaps his concept of "psychosurgery" was the medical breakthrough they had been waiting for.

Despite initial skepticism and vigorous debate about boring holes in the skull and cutting delicate brain tissue, one American physician in particular was taken by the Portuguese doctor's methodology and results. Dr. Walter Freeman, a neurologist at George Washington University, was fascinated with the idea that individual peculiarities, bizarre notions, and unattractive behaviors could be excised by simply cutting the connecting fibers between the thalamus and the prefrontal lobes. By 1936—less than a year after Moniz had invented his surgical procedure—Freeman had brought the radical technique to American shores. Freeman considered Moniz "a scientific Renaissance man impervious to criticism, whose strikingly original arcs of thought would bring him international acclaim despite his location outside the world's most prestigious centers of neurological research and learning."[17]

With Freeman as the intellectual driving force and his partner, George Washington University surgeon James W. Watts, as the physician holding the scalpel, they would transform lobotomy into one of the most controversial types of human surgery of the last 200 years.

The procedure would amplify Freeman's medical and public profile to the point that one biographer argued, "Aside from the Nazi doctor Josef Mengele, Walter Freeman ranks as the most scorned physician of the twentieth century."[18] And though Moniz was lobotomy's creator—and would ultimately receive a Nobel Prize for his achievement—it was Freeman who would travel, ice pick in hand, from institution to institution, aggressively advocating the surgery's unique attributes and sterling triumphs. During the 1940s, 1950s, and 1960s, lobotomy would become a key and often-used treatment by psychiatrists from Europe to America and Africa to Asia. In hindsight, the procedure's quick adoption by Moniz's and Freeman's acolytes and its widespread use for over a generation underscore a sad example of the psychiatric profession's irresponsibility at a time when prudence and sound judgment were most in need.

Freeman was both committed and indefatigable—a potentially deadly combination of characteristics when possessed by a man with a knifelike instrument in his hand who frequently cut into the brains of individuals with little more than anecdotal tales of success to guide him. Freeman once wrote an ode titled "The Religion of Science" that included these revealing lines: "When a man can stand on his feet and say to his fellow man 'I know' and say it with the conviction that comes from observation, can defend his knowledge by a statement of actual happenings or conditions, then his faith in that proposition is unshakable."[19]

Freeman would soon come to believe that "lobotomy, instead of being the last resort in therapy, is often the starting point in effective treatment."[20] That religious-like fervor combined with his eager attitude was prodigiously productive. During one three-week period in midsummer 1952, for example, Freeman operated on 228 West Virginia patients in institutions throughout the state. On one day alone he operated on twenty-five female patients. Freeman's own daughter was so impressed with her father's mass production capabilities that she began to refer to him as the "Henry Ford of Psychiatry."[21] Even Watts was in awe of his partner's peripatetic nature. "The worst thing about Walter Freeman," said Watts, "was that he never got tired. . . . That was what made him the most difficult."[22]

Freeman's decision to abandon hospital surgery made his assembly-line schedule possible. He believed that he was the equal of any surgeon, that anesthesia could be replaced with ECT, and that nurses, sterile garb, and the other accoutrements of contemporary surgery only cluttered up and needlessly extended the length of the procedure. Contributing to even greater efficiency in time and money was Freeman's decision to adopt a new, revolutionary procedure: the transorbital lobotomy. Freeman believed he could more easily cut through the neural pathways between the frontal lobes and the thalamus through the eye sockets with an ice pick–like instrument. The new technique could accomplish in seven minutes what the standard Freeman-Watts lobotomy required hours to do. All he needed now was an ice pick, a portable ECT machine, and an automobile to get him from site to site.

By the late 1940s, Freeman was traveling the length and breadth of the United States performing transorbital lobotomies like a traveling salesman. One critic actually referred to him as a "one-man medicine show traveling across the continent."[23] His travels and the procedure's growing popularity alarmed more reasonable members of the psychiatric community. They believed that too many lobotomies were taking place, that preliminary psychiatric examinations were being discarded, and that postoperative analyses did not support the claims of advocates.

Dr. Nolan Lewis, the director of the New York State Psychiatric Institute of Columbia-Presbyterian Medical Center, was one of the more outspoken opponents of lobotomy. "The patients become rather childlike," said Lewis in 1949. "They act like they have been hit on the head with a club and are as dull as blazes. It does disturb me that so many lobotomies are given without any psychiatric control . . . and it does disturb me to see the number of zombies that these operations turn out. I would guess that lobotomies all over the world have caused more mental invalids than they've cured. . . . I think it should be stopped before we dement too large a segment of the population."[24]

Even Freeman's friends became alarmed at the stories they were hearing. "What are these terrible things I hear about you doing lobotomies in your office with an ice pick?" wrote John Fulton, a respected Yale researcher. "I have just been to California and Minnesota and heard about it in both places. Why not use a shot gun? It would be quicker!"[25]

It was Fulton who was out of step, however. The scientific community had embraced lobotomy as an effective and fashionable way to deal with a wide range of mental diseases. That was confirmed when the Nobel Prize was awarded to Egas Moniz in 1949. Over the course of the next four years, 20,000 Americans would undergo lobotomies, at least a third of them of the transorbital variety. And at least half of the public psychiatric institutions in the United States were now performing psychosurgery. Though knowledge of the intricacies of the brain and the workings of the mind was still in its infancy, cutting into the brain had become de rigueur for doctors and institutions that wanted to be seen as up-to-date and competent in the latest techniques to combat phobias, manic depression, and schizophrenia. Even children had become fair game for brain surgeons.

Freeman attempted his first lobotomy on a child in 1939. The nine-year-old suffered terrible temper tantrums and symptoms of schizophrenia, but the severing of neural pathways did not end well; the child was forced to return to a psychiatric hospital. Freeman next tackled even younger children: a four-year-old boy and a six-year-old girl. The girl, who had "stopped talking, tore her clothes, smashed dolls, and used toys as weapons," had been diagnosed with encephalitis by her family physician, but Freeman and Watts thought her bizarre behavior was due to childhood schizophrenia. They performed a lobotomy in August 1944 only to conclude that her "chewing her clothing and fingers, incontinen[ce], and sitting alone gazing into space [with] no affection for anybody" was a sign of relapse and the need for a second surgical procedure, which was performed. Such failures, not to mention Freeman's liberal diagnosis of childhood schizophrenia, "dogged him throughout his career."[26]

Walter Freeman had definite thoughts on the cause of seriously troubled children and how prefrontal lobotomy could help them. He believed schizophrenia in childhood was a form of psychosis that would progress until it destroyed the individual's personality. "Seclusiveness," according to Freeman, characterized a slide into schizophrenia, and this critical emotional milestone could manifest itself "before the age of two." In fact, Freeman claimed that he could detect an infant turning away from other children at the age of nine months. Not long thereafter, he warned all who took the time to listen, the child in question would "stop playing with toys" and use them as "weapons or as objects to be torn apart or otherwise destroyed." From that point there would be a direct path to increasingly serious problems; the child would reject both "praise and blame," lack "warmth or affection," and display "sullen hostility and exaggerated tantrums."[27]

Psychotic children, argued Freeman, were "ritualistic . . . to an extreme" in regard to their "playthings, clothing, washing"; their "personal habits disorganized"; and feelings like "happiness" did not exist in their world. Facial expressions usually bore this out. Such disturbed children had "a dreamy, far-off look, a sullen . . . attitude, and self-absorbed appearance." In addition, they were prone to "sluggish[ness] to the point of immobility," were generally "poor sleepers," and were subject to periods of "prolonged wakefulness and fantasy."[28]

Freeman believed he had an answer for the prayers of concerned parents. If lobotomy could produce improvements in adults with serious mental problems, it could do the same for children. By 1947, Freeman and Watts had performed hundreds of prefrontal lobotomies on a vast array of patients including "11 individuals whose schizophrenia developed before the age of ten." Some had been institutionalized, and others were still at home in the care of families that "had not sent their loved one off as a matter of convenience." Freeman believed that lobotomies would be instrumental in terminating the child's fantasy life and reducing the expenditure of emotional energy. More importantly, he argued, "the aim has been to smash the world of fantasy in which these children are becoming more and more submerged."[29]

Freeman had no hesitation about applying a radical methodology that seriously challenged adults to formative young minds. In fact, he seemed to argue that the younger the diseased mind, the more destruction the surgeon's scalpel—or leucotome, as the instrument came to be called—needed to do. As he stated, "a great deal of frontal lobe tissue has to be sacrificed in children in order to obtain any more than temporary improvement. The younger the age at which the psychosis has begun, the more posteriorly the incisions have to be made, with resulting greater disability." Besides, Freeman went on to argue, "The loss of frontal lobe tissue is better tolerated by children than by adults."[30]

Comments such as these should have signaled to Freeman's compatriots in the neurological community that he was on thin ice both medically and factually, but he and others carried on unimpeded. Freeman and Watts even admitted, "Our experience in the surgical treatment of childhood schizophrenia had been rather disappointing." Two children "died shortly after the operation," and one had undergone "three operations" and was now hospitalized with little hope of "relieving the disturbed behavior." As the doctors were forced to admit, none of their patients was capable of functioning like normal adults. They were now rather "childlike and dependent on institutions for their survival."

Nevertheless, Freeman and Watts remained wedded to lobotomy as a viable answer to everything from anxiety and manic depression to schizophrenia. "In view of the wretched prognosis in all these cases without operation," they went on to argue, "modest results indicated in the case reports have encouraged us to continue with prefrontal lobotomy in certain patients whose schizophrenia has developed in early childhood."[31]

The grim results occasioned Freeman and Watts to sometimes double down on their bets. Several children had the outrageous misfortune of being subjected to multiple lobotomies. One child had endured two operations by the time he was six; a seventeen-year-old child had received three lobotomies. The doctors' own unadorned results for some young patients seemed to confirm the devastating effects of invasive brain surgery. Some succinct postoperative reports read as follows: "No change—institution" for one nine-year-old; "profound inertia after second operation" for a fourteen-year-old; and "operative death" for a twelve-year-old.[32] It was bad enough that Freeman, Watts, and the other lobotomists surgically induced childhood in thousands of adults; it was all the more troubling that they were willing to ensure permanent childhood in so many preadolescents who were caught up in their research dragnet.

Even fourteen years later, when the controversial procedure had precipitously declined in popularity, Freeman and Watts were still resolute champions of psychosurgery. In a 1961 letter to a skeptical administrator of a California hospital, James Watts wrote, "Lobotomy is an ethical and recognized method of treatment of emotionally disturbed children." Watts, a graduate of the University of Virginia Medical School—a known hothouse of eugenic thought at the time of his matriculation—went on to write that the "operation [is] most effective in reducing hostility, aggressiveness and destructive tendencies."[33] No surprise there. If one carves through enough gray matter, one is left with a rather inert, mute, stupefied half-human—something that hospital attendants, caretakers, and even some parents might have found a decided improvement.

By the late 1950s and early 1960s, Thorazine and other antipsychotic medications had come on the scene. Acting like chemical lobotomies, they tended to relax patients, calm them down, and even eliminate delusions and hallucinations. Doctors who used these drugs no longer needed as many straitjackets, locked cells, and psychosurgical procedures to end the screaming, crying, and wild ravings that once dominated the psychiatric wards. Some inmates referred to the new drug as "brake fluid."[34]

Walter Freeman, however, refused to mothball the tools of his magnificent obsession. He continued to travel, preaching the gospel according to Moniz and performing lobotomies. In 1961, for example, "he performed a series of transorbital lobotomies on seven adolescents at the Langley Porter Clinic" in San Francisco.[35]

At one of his Langley Porter presentations in 1961, Freeman proudly brought several of his transorbital lobotomy subjects with him for a show-and-tell session. The three children he had performed surgery on included Richard, sixteen, Ann, fourteen, and Howard, just twelve. As Howard Dully writes in *My Lobotomy*, his graphic account of his unnecessary psychosurgery, "When Freeman said I had just turned twelve, the doctors were shocked. Only twelve? It was outrageous. The doctors started shouting and yelling. Freeman shouted back. Soon the whole place seemed out of control."[36]

Dr. Freeman was no shrinking violet and shouted back at his accusers but was eventually "booed off the stage." The tide had turned on risky surgical procedures like prefrontal lobotomy, but not soon enough for children like Howard Dully. He was said to have been a "normal, happy baby" at birth, but his new stepmother considered Howard a problem. They did not get along; they fought often, and she believed he stole items from the house. Everything about him frustrated her. From her standpoint, he was just "impossible to control." She took him to a series of psychiatrists, but the sessions proved unsatisfactory; they considered Howard's behavior "normal." She was then referred to "a doctor named Walter Freeman."

Freeman listened to Mrs. Dully's list of grievances and thought them "sufficiently impressive." The more he heard, the more he came to believe the boy had "childhood schizophrenia." Meetings with Howard's father and four with Howard himself went reasonably well. A careful reading of Freeman's notes would suggest that nothing more than counseling was in order, but Mrs. Dully was determined that something more drastic be done. Eventually, Freeman acceded to her wishes and wrote in his notes, "The family should consider the possibility of changing Howard's personality by means of transorbital lobotomy."[37] The decision was made on Howard's twelfth birthday.

The boy was taken to a small general hospital in San Jose where the surgery was performed. He came out of the relatively short procedure—which included an electric shock to knock him out and a prefrontal lobotomy—with black eyes, a severe headache, a stiff neck, acute sluggishness, and a 102.4-degree fever. Freeman prescribed a "spinal puncture" and hefty doses of penicillin as postoperative therapy. Howard's heartrending return home was captured by his brother Brian, who would later write, "You were sitting up in bed, with two black eyes. You looked listless. And sad. Like a zombie. It's not a nice word to use, but it's the only word to use. You were zoned out and staring. I was in shock. And sad. It was just terribly sad."[38]

Howard Dully now wonders how his life would have been different if he hadn't come in contact with Dr. Walter Freeman. He also wonders, "Where were the authorities? Freeman wasn't a licensed psychiatrist. How could he determine on the basis of a couple visits with me that I had been schizophrenic since the age of four? And why would anyone accept his diagnosis anyway without insisting that I be seen by someone with the proper training? Was there no medical standard for giving someone a lobotomy, especially a child?"[39]

Apparently not.

In his history of psychiatry during the twentieth century, Edward Shorter argues that, "In retrospect, frontal lobotomy was indefensible for ethical reasons. . . . While results were dramatic, many of these patients may have sooner or later spontaneously recovered. And the irreversible damage to their brain and spirit must be weighed against the months or years with which they would have encumbered the institutional system."[40]

What Shorter could have added to his summary of an era's attraction to extreme remedial measures is that as bad as it was for tens of thousands of

American citizens to be lobotomized, given the medical community's scant knowledge of the complexities of the brain and the fact that the procedure was irreversible, the profession's decision to subject young children to the same draconian crucible magnified the collective error in judgment.

Psychosurgery would forever after have a particularly ominous and chilling aura. As doctors learned more about the intricate workings of the mind, they seemed less willing to disturb delicate brain tissue, which led to fewer instances of invasive brain surgery. Popular novels such as Ken Kesey's *One Flew Over the Cuckoo's Nest* only added more baggage to an already suspect medical procedure. A few doctors, however, weren't intimidated by the negative newspaper articles, salacious magazine pieces, and growing patients' rights movement that opposed such nefarious undertakings.

Dr. O. J. Andy was one of them. The director of neurosurgery at the University of Mississippi, he continued to perform a wide variety of brain surgeries into the 1970s and, like Walter Freeman, found it equally acceptable to perform such surgery on children. Andy was interested in treating abnormal behavior, those behaviors that did not respond to psychotherapy or medication, including emotional instability, aggression, hyperactivity, and nervousness. All of these symptoms of abnormality, according to Andy, had "become manifest in children."[41]

And like Walter Freeman before him, he saw little reason to wait until the patient reached adulthood to address the symptoms. W. B., for example, was a twelve-year-old "mental defective with seizures, repetitive movement and behavior disorder." Andy believed a thalamotomy (a procedure where a portion of the thalamus is destroyed) was in order and performed one on the left side in February 13, 1964, and on the right, later that same year on December 17, 1964. Andy considered the surgery a success: "the rocking, self-beating, regurgitation and destructiveness ha[d] ceased" . . . and the boy was "up and around with other mentally retarded patients."

Another boy to receive the procedure was J. M., a nine-year-old who "had seizures and behavioral disorder" that included "combative" and "destructive" behavior. Andy performed a "bilateral thalamotomy" on the left side in January 1962 and a similar procedure on the right in September only to witness many of the negative behaviors return a year later. A fornicotomy (surgical destruction of a bundle of fibers in the brain that carry signals from the hippocampus to the hypothalamus) was then performed in January 1965, resulting in the boy's having an "impaired memory" and greater "irritability and combativeness." Undaunted, Andy dug into the child's brain once again, this time performing a "simultaneous bilateral thalamotomy." He thought this fifth operation a success; the child had "become adjusted to his environment and displayed marked improvement in behavior and memory." Less optimistically, however, Andy admitted that the patient was "deteriorating intellectually."[42]

After so many invasive procedures to destroy brain tissue in such a short period of time, it's a wonder the child could function at all. The one discussant to the journal article offered comments regarding several technical issues but made no mention of Andy's inclusion of children in such irreversible surgery. Nor did the article make any mention of parental permission to conduct the surgery.

One physician at the time was willing to speak out about the use of children in such devastating surgery and Andy's cavalier history of repeatedly performing such surgery. Dr. Peter R. Breggin was a reform-minded psychiatrist who was active in national antipsychosurgery campaigns. He believed there might have been a political agenda behind Andy's research, which explained the inordinate number of African Americans and criminals used as psychosurgery subjects.

Breggin began researching the resurgence in psychosurgery in the early 1970s and came upon the work of O. J. Andy. Andy had published surgical reports on approximately three dozen young children, ages five to twelve, who were diagnosed as aggressive and hyperactive. Breggin contacted a civil rights attorney in Mississippi who was able to determine that most of the children operated on were housed in a segregated black institution for the developmentally disabled. Nurses told the attorney that Andy had a completely free hand in picking children for psychosurgery.[43]

Breggin would go on to criticize Andy's work when he testified in Washington at congressional hearings on "human experimentation" in the aftermath of the Tuskegee syphilis study revelations. In fact, Andy was there to testify as well, along with a number of the nation's top surgeons and medical administrators. Senator Edward Kennedy, who chaired the hearing, asked Andy a series of questions, including how many patients he had practiced psychosurgery on, what their ages were, and when he had started performing such surgery.

Andy admitted to doing brain surgery in the early 1950s on "30 or 40 patients" of which "13 or 14" were children. He claimed the youngest were between "6 or 7 years of age."[44]

Where Andy was the model of brevity when questioned, Breggin was far more expansive in his comments. Obviously dissatisfied with the laissez-faire approach to medical research, Breggin said, "The reliance on professional ethics and medical control over these issues leaves the physicians in charge of the situation. It creates for themselves an elitist power over human mind and spirit." Under questioning from Kennedy, Breggin went on to declare his opposition to psychosurgery, a procedure that "destroy[s] normal brain tissue." The practice "opens up a Pandora's box of possible approaches; men like Dr. Andy invent diseases to operate on."

In addition, said Breggin, there was no oversight or peer review. He claimed that when he asked Andy's superior at the University of Mississippi, the chairman of the Department of Psychiatry, whether he knew Andy was doing surgery on children, the man replied, "Oh, my God, no."[45] Breggin went on to castigate both the procedure and the doctor performing it.

Decades later, psychiatrists and medical ethicists admit that doctors still do not know much about the circuits they are tampering with or what will result from their intervention: "Some people improve, others feel little or nothing, and an unlucky few actually get worse."[46] Institutions today conduct ethical screening to select candidates, insist that potential subjects must have disabling maladies, and disclose in informed consent instructions that the operation is experimental. Surgical precision is emphasized today, but "there still remain large gaps in doctors' understanding of the circuits they are operating on." That surgeons—and, in Walter Freeman's case, nonsurgeons—were wielding scalpels and icepicks and "slashing into the brain" during frontal lobotomies on thousands of patients and "blindly mangling whatever connections and circuits were in the way" is a perfect example of the well intended doing great harm. That they chose to perform this risky procedure on children only magnifies their flawed judgment and self-serving approach to research and medicine.

**"FACED WITH PROFOUND CLINICAL PROBLEMS,"** writes medical historian Joel Braslow, "doctors saw lobotomy as a humane solution."[47] From the perspective of frustrated and overworked doctors, the situation is not difficult to understand. The number of asylums was increasing, yet psychiatric wards were overflowing, and effective and lasting remedies were few and far between. Doctors confronted with a laundry list of mental diseases and inexhaustible caseloads remained desperate for cures, as did patients themselves. Any new theory, technique, or drug that showed promise sparked interest. Some would become wildly enthusiastic promoters of a particular treatment regimen and forever be closely associated with it—perhaps not to the extent of the association between Walter Freeman and the transorbital lobotomy, but devout champions of a particular elixir or procedure. And Lauretta Bender was one of them.

As mentioned, early in her psychiatric career, Bender had believed shock therapy was an effective treatment for a whole range of mental problems, even for very young children. Initially at Bellevue and later at Creedmore State Hospital in Queens, she assigned a childhood schizophrenia diagnosis to an inordinate number of young patients. A regimen of twenty shock-therapy sessions would usually follow. For children like Ted Chabasinski, who were troubled but not schizophrenic, the ECT sessions were more punitive than therapeutic.

Two decades later, Bender would discover a new medicinal elixir—LSD—and utilize it almost as aggressively. As a zealous enthusiast of the controversial hallucinogen, she saw no bounds in its potential application.[48]

In 1943 Dr. Albert Hofmann, a Swiss chemist for Sandoz Pharmaceutical Company, accidentally ingested a chemical he had been working with in the laboratory. He quickly noticed dramatic physiological and psychological changes; it was his first psychedelic trip. The substance he was working with was an ergot fungus, and its bizarre hallucinatory impact captured his curiosity. He called the chemical derivative lysergic acid diethylamide 25 (LSD), and it would quickly become the most sought-after and controversial hallucinogen in the world. Within a few short years, many believed LSD was the long-sought potion that held the keys to unlocking the universe.

During the war, Nazi doctors at Dachau and other concentrations camps used prisoners for a host of experiments, including mescaline studies. They were searching for mind-control techniques, or incapacitants, that would help immobilize and defeat an enemy. American authorities also were exploring new chemical concoctions that would have wartime utility. However, they had not progressed much past truth serums using various forms of marijuana. Hofmann's psychedelic creation was considerably more potent. Interest in

LSD was enhanced during the early years of the Cold War, with numerous, inexplicable incidents of soldiers, citizens, and political leaders in Europe and Asia confessing to crimes they could not have committed. Cardinal József Mindszenty in Hungary, captured American troops in Korea, and several other disturbing examples of individuals under someone else's control alarmed American leaders. The new threat of mind control and brainwashing gave everyone a fright and exacerbated fears of the Soviet Union and the rise of Communism.

Members of the defense establishment, particularly those in the newly established Central Intelligence Agency (CIA), quickly recognized that mind control had both defensive and offensive capabilities. "Nearly every Agency document," according to author and former State Department official John Marks, "stressed goals like controlling an individual to the point where he will do our bidding against his will and even against such fundamental laws of nature as self-preservation."[49] In 1950, the CIA initiated a top-secret plan called BLUEBIRD to monitor Soviet developments in the behavioral field and explore America's potential regarding mind control. A year later it would be renamed ARTICHOKE and the following year rolled into MKULTRA. Those and the many other incarnations that followed were secret government programs designed to investigate everything from novel chemical substances like LSD to sensory deprivation, brain electrode implants, and hypnosis. Human experimentation would be the methodology that determined which of these clandestine efforts showed the most promise. "We lived in a never-never land," said one CIA doctor, "of unceasing experimentation."[50]

Techniques as varied as ultrasonics, high and low pressure, poison gases, radiation, extremes of heat and cold, and changing light were all being explored. Marks even discovered evidence that in 1952 the Office of Scientific Intelligence considered giving a doctor $100,000 to study "neurosurgical techniques," which probably meant lobotomy and electric shock.[51] The memory of Hitler was still fresh, Stalin had taken his place in the pantheon of evildoers, and the global threat of Communism was increasingly apparent. Apologists for the scientific excesses that were about to be committed quickly underscore "the fear—even paranoia" that gripped the American political and social landscape. Just as World War II provided an excuse for various ethical transgressions in the medical arena, the toxic Cold War atmosphere of the 1950s and 1960s provided cover for similar excesses. Some embarrassing episodes were kept secret for decades so as not to expose America's own questionable, if not criminal, behavior. The deaths of Harold Blauer, a former professional tennis player, and Frank Olson, a biological weapons expert, in 1952 were caused by dedicated but reckless government operatives pursuing greater knowledge about hallucinogens and mind control. Equally repugnant, though no loss of life was reported to have occurred, was the CIA's funding of respected scientists in their exploration of various and occasionally crackpot mind-control methodologies. One of the stranger cases was that of Dr. Ewen Cameron, an esteemed Scottish-born psychiatrist at Montreal's McGill University, who used electroshock, drugs, psychic driving, and an assortment of other bizarre methodologies to depattern patients and restore them to a "desired" condition.

A coterie of accomplished psychologists and psychiatrists signed on as covert scientific collaborators with the CIA in exploring LSD's full potential.[52] Other doctors and academics had no formal ties to the CIA, but their research was of great interest to the agency. As some observers now believe, "it was impossible for an LSD researcher not to rub shoulders with the espionage establishment, for the CIA was monitoring the entire scene."[53] Some of these investigators during the 1950s and 1960s claimed that LSD had a positive effect on everything from alcoholism to homosexuality. One believed the new hallucinogen was capable of making revolutionary inroads in fighting childhood schizophrenia and autism.

As an enthusiast of new groundbreaking psychological techniques, electronic gadgets, and drugs that showed particular promise, Dr. Lauretta Bender knew how to discard caution and skirt codes of conduct to satisfy her scientific curiosity. There is no better example of this than her dubious decision to give dozens of young autistic children daily doses of LSD. During the 1950s, Bender began experimenting on children with a number of tranquilizers and psychopharmacological agents. In one journal article on the subject, she tipped her philosophical hand with the off-the-cuff comment, "I am one of the old clinicians who feel that research cannot be done by pattern, it has to be done by inspiration."[54]

The inspired notion of breaking with convention and prudent decision making and giving children the most controversial psychoactive drug under investigation must have given pause to the more conservative members of the psychiatric community. The idea "that LSD could be used to *treat* psychological problems seemed downright absurd to certain scientists in light of the drug's long-standing identification with the *simulation* of mental illness."[55] Bender, however, thought otherwise.

In 1960 Bender began giving LSD and UML–491, a derivative of LSD, to children at Creedmore State Hospital. There were initially fourteen subjects between the ages of six and ten years old, though the number of participants would quickly increase. Eleven were boys, three were girls, and all were said to be either schizophrenic or autistic. Because Bender's diagnostic criteria were considerably different from those today, it is extremely difficult to determine the children's actual disorders. Initially the children were given small doses that were gradually increased and divided into two daily doses. Incredibly, some children received this regimen—hefty even for adults—for a year or more.

Bender's comments about the experiment in a subsequent journal article are quite revealing. After being given the LSD, a number of formerly quiet children became quite "aggressive—pushing, biting, or pinching other children." Bender considered this an improvement over their past behavior, in which they basically ignored the surrounding environment.[56] And if the LSD was not enough to break through each child's defenses, Bender planned on adding ECT or insulin therapy "to take maximum advantage of the intensified" stimuli—a physical and mental battering by any standard of measurement. Even more revealing was this comment: "Because of the extremely violent psychotic reactions described when adults were given large dosages of LSD, we were extremely cautious when first using the drug, even obtaining parents' consent."[57] The implication of such an admission, of course, is the probability that Bender routinely incorporated institutionalized children into research studies without first asking their parents' permission.

During the course of a conference presentation and a subsequent journal article reviewing her use of Metrazol and ECT on children over the years, Bender made this rather strange statement:

> Now we have not determined who the sinner may be. There are many who will believe that the mother is always the sinner; is she not schizophrenic? There are others who will believe that the individual himself is the sinner since, after all, is he not the one whose behaviors and fantasies are schizophrenic? However, there will be many of you attending this fifty-second annual meeting of the American Psychopathological Association, who will

judge the speaker to be the sinner, for have I not given you this wealth of clinical material without charts, graphs, or statistical evaluation?[58]

There is no indication of how those in the audience reacted to this odd remark.

In a journal article dealing with her Creedmore LSD experiments, Bender admitted that some children—both autistic and nonautistic schizophrenic boys between six and twelve years old—were kept on a two-dose-a-day LSD regimen for weeks, months, and in some cases a year or two. Though most prudent physicians would, no doubt, shun such a dangerous experimental venture, Bender was pleased with the results: "These were mute, autistic, and schizophrenic children. And they showed a general improvement in well-being, appearance, and lift in mood."[59]

In one rather ironic admission, Bender said that two adolescent boys in the study "became disturbed to the extent that they said we were experimenting on them." Months later, one of these boys complained once again that "we were experimenting on him with the drug and trying to keep him from getting out of the hospital." Bender said they soon after decided to end the boy's participation in the study "not because of the drug but because of the boy's attitude toward it, based on his own psychopathology."[60] The boy's awareness and sense of self-preservation lead one to believe he may not have been as dysfunctional as Bender believed.

It would be in this same 1970 article that she admitted using "LSD on 89 children from January 1961 to July 1965 . . . because we have found that it is one of the most effective methods of treatment we have for childhood schizophrenia." Also in this article Bender admitted keeping the children isolated as they went through the drug reaction—something that doctors even then would have absolutely opposed in both theory and practice.[61]

Almost a half-century later, the vast majority of Americans—including most doctors—are unaware that a noted child psychiatrist conducted such unethical and dangerous psychotropic medical experiments on children. Whether it was due to Dr. Bender's firm conviction that LSD held the keys to unlocking the mind's secrets, to her desire to become the next "god of science," or just to her desire to accumulate knowledge, the research went on without comment or opposition. This is a sad commentary on the way in which concerted action by powerful people can cow intelligent, sophisticated members of an esteemed profession. Clearly, Lauretta Bender was not the only person to provide LSD tablets and injections to scores of children over the months and years the research project was in operation. Why did no one in authority speak up? Why didn't another doctor or nurse or medical administrator say stop, this is wrong? Did no one recognize the harm that was being done?

Walter Freeman had a dangerous fascination with performing lobotomies on as many "mentally ill" people as he could; Bender was also a passionate propagandist of illusory nostrums. As an avid enthusiast—first with shock therapy in the 1930s and then with powerful hallucinogens twenty-five years later—she was convinced that she had the answer to a series of long-troubling mental mysteries. That she repeatedly trampled on the rights of patients, violated professional codes of conduct, and carried out her experiments on society's most vulnerable members and was still honored by her colleagues speaks volumes about the medical profession at the time.

# NINE

# PSYCHOLOGICAL ABUSE

## *"I Call That Brainwashing"*

*They should have realized what they were doing before they started the experiment. They were educated. They weren't stupid. They destroyed people's lives for science, and then they covered it up. Now they're sorry because they got caught.*

—Mary Korlaske Nixon

**WENDELL JOHNSON ARRIVED ON THE CAMPUS OF THE UNI-**
versity of Iowa in 1926 with great expectations. With Sinclair Lewis's *Arrowsmith* and Paul de Kruif's *Microbe Hunters* garnering national recognition at the time, good writing and important scientific research were helping to make serious scholarship a promising and exciting venture. Johnson was a fine athlete and a good student. He set his sights on accomplishing two goals in college: becoming a writer of consequence and improving his speech.

For most of his childhood and adolescence, Johnson had been plagued with a severe stammer. The affliction had been a profound embarrassment that often led to teasing, name calling, and impromptu fisticuffs. Johnson even picked up the nickname Jack, after the heavyweight champ Jack Johnson, for his propensity to punch out classmates with the temerity to tease him about his stuttering. The University of Iowa was considered a leader in the new field of speech pathology, and Johnson often "offered himself up" as a willing test subject. "In the clinic, Johnson was hypnotized, psychoanalyzed, prodded with electrodes, and told to sit in cold water to have his tremors recorded." He even put pebbles in his mouth and placed his right arm in a cast to "equalize the imbalance of the hemispheres of his brain" that some believed led to stuttering. None of the nostrums had any long-term success. As Johnson would write of his career as a test subject in his journal ten years after arriving at Iowa, "I'm a professional white rat."[1]

Johnson's preoccupation with conquering his own dilemma enabled him to become an authority on speech pathology, and he gave presentations on the subject throughout the Midwest. He could talk authoritatively on the social stigma, the embarrassment, the loss of self-respect, and the "self-imposed isolation" that stutterers often faced. Thoughts of suicide are not uncommon among stutterers. As he wrote in his journal, "The stuttering child is a crippled child."

After years of contemplating his own problem and the flawed speech patterns of others, by the mid-1930s Johnson had begun to formulate his own ideas about the origin of stuttering and the best methods for conquering it. The dominant theory at the time was biological; stuttering was the result of a genetic defect. The more Johnson thought about his own history and that of other individuals he interviewed, the more he realized that every person "had been labeled a stutterer at a very early age." As he would enter in his journal, "Stuttering begins in the ear of the listener, not in the mouth of the child."

Looking back at his own unhappy history with speech, he believed that a first-grade teacher had exacerbated his relatively minor problem with her frequent and aggressive efforts to correct him. Her recommendation that his parents do the same only compounded the problem. The more he thought about his own case and that of others, the more Johnson came to believe that "the affliction is caused by the diagnosis." If left alone, Johnson believed, children would naturally overcome their stuttering problem. This was a revolutionary notion; it was also unsupported by any scientific data. If he hoped to confirm his "diagnosogenic theory," a term he coined, he would have to formulate and carry out an original research project that would require repeated access to children in a controlled environment. On becoming a speech professor at the university himself, it would be one of the things he'd place on his research agenda.

The perfect test site was just an hour's drive from the campus: the Iowa Soldiers' Orphans' Home. There was precedent for his choice: The orphanage had been used previously for research by college professors. In fact, one of Johnson's former students, a professor of speech pathology said, "They used that orphanage as a laboratory rat colony."[2] Johnson received permission to carry out his study during the fall semester of 1938. The last piece of the puzzle fell into place when he asked Mary Tudor, a graduate student, "Have you chosen anything for your master's thesis?" Her negative reply would allow Johnson to suggest one. And though that meeting would enable Tudor to get her degree and help confirm Johnson's theory, it would also culminate in the most embarrassing and regretful chapter of their professional lives.

Essentially, Tudor was told that she would be working with two groups of orphans from the Iowa Soldiers' Orphans' Home: one of stutterers and another of normal speakers. Half the children would be assigned to an experimental group, the other half to a control group. Children in the control groups would be labeled normal speakers and receive positive therapy. Children in the experimental groups would be labeled stutterers and given negative therapy.[3]

It was imperative that Tudor point out stuttering among children in the experimental group (even when none existed) so that they would be sensitized to their speech and made conscious of their speaking habits. Tudor was to sternly lecture the children when they repeated a word, mispronounced it, or even stopped midsentence. Speaking, in essence, was to be made an ordeal. If Johnson's theory proved correct, they would have created a group of young stutterers.

Tudor was told she would have to lie to the teachers and matrons of the orphanage and tell them she was there to do speech therapy. They were to follow her orders in correcting the children when she wasn't on the premises. She had no qualms about the experiment. Johnson was a rising star in the field, and any association with him was bound to help her career.

Tudor found the Iowa Soldiers' Orphans' Home a depressing place. She imagined how unfortunate it would be to grow up in such a cold, austere environment, but Johnson's theory made sense to her, and it was flattering to be asked to be part of such an important project of which she would be the sole author. After going through files and examining "the speech of 256 orphans, she and the other speech pathologists culled 22 subjects: 10 stutterers and 12 normal speakers. They paired the children based on similarities in age, sex, IQ, and fluency. Then they randomly assigned one from each pair to the control group and the other to the experimental group." Just that simply, Tudor and Johnson would orchestrate a human research project that would prove psychologically destructive to the lives of young children. Many decades later it came to be known as the infamous Monster Study.

As one of the luckier children would say after learning what had happened to her friends some sixty years earlier, "There but for the grace of God, I could have been placed in an experimental group. It could have been my life that was destroyed."

Initially, children at the orphanage were delighted to be part of such a venture, much as the Fernald boys were when told of the new Science Club; like the Fernald boys, they had no idea what was in store for them. Visitors were rare, and to be the center of attention was unprecedented. Mary Korlaske, for example, was a twelve-year-old with an IQ of 81 who had been in the orphanage for five years.[4] The Depression had wreaked havoc on her family; her mother had sent her and two older brothers off when she was seven. On her arrival home one day, her mother quickly ushered her into a waiting black sedan with no explanation other than "You'll be safe. You'll be all right." Tender, loving care were certainly not watchwords at the Davenport institution.

Young Mary Korlaske was smitten with the tall, slender college student who was showering her with attention. She wondered if Mary Tudor, just twenty-three at the time, was there to adopt her and become her new mother.

Mary Korlaske's first interview was on January 19, 1939. Tudor asked her if she knew anyone who stuttered. As she answered that and other questions, Tudor frequently interrupted her and told her she was beginning to stutter. She went on to sternly warn Mary that if she didn't address the problem quickly, she would go through life with a pronounced stutter. Tudor would write in her notes of that day that Mary Korlaske "reacted to the suggestion immediately and her repetitions in speech were more frequent."

Tudor offered the young girl, who was enamored with her, suggestions to combat her stuttering, but they were really psychological trapdoors and tricks: "negative therapy" designed to make Mary more self-conscious about her speech. "Take a deep breath before you say the word if you think you're going to stutter on it," advised Tudor. "Stop and start over if you stutter. Put your tongue on the roof of your mouth. Don't speak unless you can speak correctly. Watch your speech all the time. Do anything to keep from stuttering." The advice was the direct opposite of what Wendell Johnson truly believed ameliorated a stutterer's dilemma.

For "Case Number 11," a five-year-old who was a normal speaker but had been declared a "stutterer" by Johnson's team, the sessions were stressful from day one. Mary Tudor had asked the young girl to "tell me the story of the three bears." When the five-year-old "repeated the word 'she' three times," Tudor "called her attention to it and told her that that was stuttering." Tudor wrote in her thesis:

> I told her to stop when she repeated words and take a deep breath, or to stop
> and begin again and try to say them just once. The next time she repeated,
> I stopped her and she reacted immediately. From then on when she had a
> repetition she stopped, put her hand to her mouth and gasped. Then she
> laughed and tried to do it over again. She became conscious of her difficulty
> immediately. She noticed her own mistakes and she began to cut her words
> off precisely.

Additional "negative therapy" sessions with the girl only worsened her speaking ability. The problem grew to the point where the little girl refused to speak. As Tudor wrote of the girl's downward spiral: "It was difficult to get her to speak although she spoke very freely the month before. I asked her why she didn't want to talk. She didn't answer. Then I asked her if she was afraid of something. She nodded her head. What are you afraid of? Tudor asked. After some time she said, 'Afraid I might stutter.' She reacted to every repetition by stopping and hanging her head. She looked down practically all of the time. She seemed inhibited and she didn't smile."[5]

Johnson's theory of stuttering was proving correct, and Tudor's graduate thesis was moving along nicely. For the children at the orphanage, however, there was less reason to be cheerful.

Tudor returned to the orphanage every week or two for additional sessions with the students. They were given verbal exercises and drills that stressed them. When Tudor was not there, matrons at the orphanage continued the barrage of criticism and destructive advice. The experimental group witnessed a steep decline that affected more than their quickly diminishing communication skills. They were made the brunt of jokes, their grades dropped, and some children began to withdraw and isolate themselves.

After several months, Tudor was also showing the effects of her emotionally exhausting work. She wasn't thrilled with her treatment of the children, but Johnson was delighted with her progress, and his enthusiasm recharged her occasionally flagging spirits. As Tudor would one day admit, "I didn't like what I was doing to those children. It was a hard, terrible thing. Today, I probably would have challenged it. Back then you did what you were told. It was an assignment. And I did it."[6]

In late May 1939, Johnson traveled with his student to gauge the final product of her work. Of the key experimental groups, five of the six once-normal speakers were now stuttering, and three of the five stutterers had further deteriorated in speaking ability. As expected, the control groups were relatively unaffected by the experiment.

Tudor spent that summer transcribing her many Dictaphone recordings and writing her thesis. Many of the pages from the 256-page document contain comments like this of "Case Number 11":

> [T]his subject showed a decrease in fluency . . . and an increase in the percent of speech interruptions. During the experimental period her manner of speaking changed markedly. At the beginning of the period she spoke freely and connectedly, but at the end of the period she was very unwilling to talk. . . . She also ceased to tell stories or converse freely with her playmates.[7]

Generally, said Tudor, "all of the children in [the experimental group] showed overt behavioral changes in the course of the experiment that were in the direction of the types of inhibitive, sensitive, embarrassed reactions shown by many adult stutterers. . . . There was a tendency for them to become less talkative." Interestingly, even though Tudor had delineated and documented how the exercise had stamped in each child's mind that "there was something definitely wrong with their speech" and "noticeably" altered their behavior, she recommended that they "repeat this study under conditions more nearly comparable to the home situation" most children live under. "More extensive results might reasonably be expected," argued Tudor, if a "home situation" were the site of future research.[8] Fortunately, there is no record of Johnson or Tudor replicating the study.

Mary Tudor would also include in her thesis some interesting comments about the staff of the Iowa Soldiers' Orphans' Home. "Both the teachers and the matrons of the institution were very easily influenced, that is, they accepted the diagnosis given without questioning." Moreover, wrote Tudor, the staff believed "children of such 'caliber' and in such environmental surroundings could not be helped. They apparently felt that it was a waste of time to give these children special attention."[9]

By the end of that summer, Tudor's hefty research document had been submitted and approved, and she was off to a job as a speech therapist in Wisconsin. Though it was over for her, those who made up the experimental groups at the Iowa Soldiers' Orphans' Home were still receiving speech "therapy" from the institution's teachers and matrons. By now Mary Korlaske and others were stuttering badly, and it was having a dramatic effect on their behavior, their schoolwork, and their relations with other children and orphanage staff. Orphanage officials contacted Dr. Johnson and expressed their concern. Could he do something to address the problem? Johnson, in turn, reached out to his former graduate student and asked if she couldn't visit the institution during her Christmas vacation and see if some positive therapy would work.

Tudor's visit was brief, depressing, and unsuccessful; the speaking ability of the children in the experimental group was significantly worse. Nothing she did improved the situation. Two additional visits proved equally futile; she could not reverse the damage she had done.

During the war, Mary Tudor worked as a procurement officer for the navy. When the war ended, she returned to Iowa with the hope that her former mentor would assist her in getting a job in some aspect of the speech therapy field. The reunion was less than cordial. "It was clear he didn't want me around," Tudor recalled. "He was worried I'd tell somebody."

Apparently, the intervening years had been instructive for Wendell Johnson. His graduate students questioned the ethics of the Iowa orphan study and began referring to it as the Monster Study. It was something that should be hidden, not celebrated. Some even compared it to what the Allies had discovered on entering Ravensbrück, Treblinka, and Dachau. The Nazi medical experiments had sensitized Americans to the misuse of test subjects, and the research project at the orphans' home could easily strike some as similar in design and impact. "This was the kind of stuff you would think they were doing in Auschwitz and this is why, at the time, people concealed it," said Franklin Silverman, a speech pathology professor and former student of Johnson's. "They wanted to block it out of their minds and make believe it didn't happen."[10]

Johnson must have been on the horns of a considerable dilemma. The research had proved his theory sound, but the project also disturbed and angered people; its ethics were corrupted from the start and had caused great damage to innocent children, orphans no less. How does one handle such a sensitive matter in which a research project confirms a groundbreaking theory and will launch its author to the highest ranks of the profession but is so explosive and repugnant to normal sensibilities that it could also cause that individual to be shunned, if not jettisoned from the academy?

As one college communications professor and former Johnson student said, "He didn't know how to react to it or handle it."[11] It is probably that quandary that prevented Johnson from sending his speech students back to the Iowa orphanage to undo what he had helped create. Rather than addressing the situation in a concerted and forthright fashion, Johnson chose to try to forget about it and hoped that everyone else would as well. In the end, he chose a prayer for collective amnesia rather than something more meaningful, such as a profile in courage.

Johnson was not about to sandbag his diagnosogenic theory, however. He shoehorned it into the literature by arguing from an anthropological perspective that some Indian tribes had no concept or word for stuttering. "The Indian children were not criticized or evaluated on the basis of their speech, no comments were made about it, no issue was made of it," Johnson argued in his 1946 book, *People and Quandaries*.[12] He made no mention of his student's research, a project that documented his important, new theory, and one that would subsequently transform speech therapy around the world.

When Johnson died in 1965, the few who knew of his secret were still hiding it. Through the audacious and controversial efforts of *San Jose Mercury* investigative reporter Jim Dyer, the Monster Study was finally revealed. His series on the Iowa Soldiers' Orphans' Home research project in the summer of 2001 attracted nationwide coverage. Mary Korlaske Nixon was seventy-five when she learned what had actually happened at the orphanage, and she was furious. After her participation in the experiment, she had been teased, laughed at, and ultimately sent to a reform school for girls. Her lawyer admitted she has "a very poor self-image. She's afraid to speak out, not very outgoing. . . . She's lived her whole life like that."[13]

Not long after, Mary Tudor, then eighty-four and living in California, received a letter from her former speech pupil, Mary Korlaske Nixon. It was one she would rather not have received. "I remember your face, how kind you were and you looked like my mother," the letter began. "But you were ther [*sic*] to destroy my life." The letter writer went on to call Tudor a "monster" and a "Nazi" for what she had done to her and the other children at the orphanage sixty years earlier.

Back in Iowa, the university was forced to apologize. Dr. David Skorton, the vice president for research, said, "This is not a study that should ever be considered defensible in any era. In no way would I ever think of defending this study. In no way. It's more than unfortunate. But this man made enormous contributions, both by his direct work with patients and with training so many practitioners and students over the years."

The former orphans brought legal action. Friends and former colleagues of Wendell Johnson were left trying to make sense of it all. "I think it's not coincidental that he chose to do it with a group of parentless kids," said Tricia Zebrowski, an assistant professor at the Wendell Johnson Speech and Hearing Center at the University of Iowa. "This was the only way he was going to get kids," she said of Johnson's decision to use institutionalized children. Though she told Dyer she didn't agree with Johnson's approach and thought the study unethical, she did not consider Johnson a bad person. "I think it was just the culture of the time." Duane Spriestersbach, a former colleague of Johnson's, also argues that "it was a different time and the values were different. Today we might disagree with what he did, but in those days it was fully within the norms of the time." Not all agree; many argue that there were rules and standards that precluded causing harm to test subjects, especially children.

In 2007, the six plaintiffs who filed suit eventually won a settlement of $925,000. Hazel Potter Dornbush, eighty-four at the time of the settlement and one of the orphans chosen for the study, found it hard to forgive the study's organizers for what they did. "I call that brainwashing. I don't care what anybody else calls it, that's my language. I was wise to it right away, but I cooperated. You know we weren't in no [*sic*] position to argue with nobody. We had nobody to lean on to help us out."[14]

**PROFESSOR WENDELL JOHNSON'S ABILITY** to stress test subjects in the course of scholarly research and divorce himself from ethical standards that he would normally advocate and abide by is, in fact, not that unusual. There is no shortage of accomplished and respected academics who breached standards of ethics when blindly pursuing a particular point of interest. Whether in the hope of confirming a new theoretical breakthrough, cementing a long-sought financial connection with a pharmaceutical company, or just trying to establish themselves as legitimate scholars and researchers, investigators took liberties they shouldn't have taken. The attraction of the all-consuming scientific hunt was only heightened—and the need to follow prescribed rules and regulations was whittled away—when outside pressures, obligations, and enticements intruded to make the calculus of everyday decision making that much more problematic.

World War II had that effect on many normally prudent scientists. The Cold War that was to follow had a similar impact, especially for those who were on the front lines of America's defense establishment. The famous Harvard psychologist Henry Murray was one of them; late in life his psychological stress research would gain him some undesired notoriety. As one commentator would write of him, "Murray was an anomaly among academic psychologists, and controversy surrounded his career."[15]

Born at the end of the nineteenth century, Murray was from a wealthy New York family and was educated at the finest private schools. He would go to Harvard and major in history, but his mediocre academic record as an undergraduate would not have predicted the grand things that were to come. Murray went on to medical school at Columbia and also earned a master's degree in biology. After teaching at Harvard for a few years, he traveled to England, where he earned a PhD at Cambridge in 1928. Murray's eclectic spirit would lead him to spend time with Carl Jung, fostering his newfound interest in psychology and psychoanalysis and increasing his curiosity about various matters of the mind.

Back at Harvard, Murray would make a name for himself exploring notions of latent needs, external influences on motivation, and the concept of "apperception." In the 1930s he would develop the Thematic Apperception Test, a psychological evaluation device that encourages individuals to tell stories based on pictures they are shown, and would write the well-known psychology text *Explorations in Personality,* both of which added gravitas to his reputation as a scholar and creative thinker.

Though increasingly accomplished, Murray was hard to peg down. "Murray was trained not in psychology, but rather in medicine and biochemistry. Furthermore," says one biographer, "Murray rejected reductionism in psychological theory and research, instead promoting methods derived from medicine and a theory of personality influenced by psychoanalysis and . . . biological systems theory."[16] The cross-pollination of disciplines was hard for the Harvard faculty and administration to digest. Fortunately, Murray had the financial backing of the Rockefeller Foundation, which allowed him to continue at the university and further his study of personality development.

During World War II, Murray's knowledge of personality traits was used by the military to select secret agents for special intelligence assignments. Much of his work was on behalf of the Office of Strategic Services (OSS), the forerunner of the Central Intelligence Agency. In addition to matching personality and character traits to military job descriptions, Murray was asked by OSS chief William Donovan to develop a personality assessment of Adolf Hitler. Murray's analysis, combined with that of several other scholars, contributed to one of the first mergers of criminal profiling and political psychology in the pursuit of character prediction. The team's prognostications, including that Hitler would commit suicide, gained them and their new profiling system additional plaudits. After the war, Murray returned to Harvard but continued work as a CIA consultant. His shadowy research work for the spy agency would shock the conscience of those in the academic community who were unaware of his clandestine ventures.

In 1958, Murray began stress research that had definite CIA and Cold War implications. The project, which at least one author believes had MKULTRA associations, was designed to establish an assessment scale that would collect an impressive amount of data about the reactions of people under pressure. Twenty-five Harvard students were selected to participate, and the aim was to assess their reactions to stress. Murray's experimental protocol measuring personal reactions to intense, sustained, and highly insulting criticism was unusual but relatively simple in design. The student subjects were told they had a month to write a brief composition regarding their individual philosophies of life. Once submitted, an experienced lawyer would confront each subject about his respective philosophy and attitudes. Both personal and general issues were considered fair game. The subjects were informed that the debate would be filmed; they were not told, however, that the attorney's criticisms would increase in number and tone and become personally abusive.

Designed to be extremely stressful, the dialogue was more a psychological endurance contest constructed to measure how much verbal abuse an individual could withstand. The students were connected to monitoring devices that charted their psychological reactions as the lawyer pointedly and repeatedly belittled aspects of their personal essay and stand on issues. Subsequently, each student was forced to watch a tape recording of the session illuminating his or her impotence and rage. Students were also told to rate their feelings about the lawyer. At two- and eight-week intervals, they were required to return to watch the tape again and answer additional questions.

One of the student subjects was a sixteen-year-old math prodigy named Theodore John Kaczynski; today he is better known as the Unabomber. An excellent student from a second-generation Polish American family, Kaczynski had always been more comfortable with numbers than with people. Growing up, he had few friends but many books. He skipped several grades in the Chicago school system and graduated from high school at fifteen. At Harvard, he excelled in mathematics and was taught by some of the nation's top mathematicians, who vouched for his exceptional ability with numbers.

According to one observer, the precocious math student was emotionally stable when he became a test subject in Murray's research project. Kaczynski's lawyers argue that some of his emotional instability is directly related to his participation in Murray's research project. They believe the young math student cracked under the pressure.

It is further argued that, over time, Kaczynski would embrace ideas that "were products of two historical trends: a crisis of reason and the impact of the Cold War."[17] Kaczynski came to believe that only empirically verifiable statements were meaningful. The Cold War accelerated this process. The threat of Communism and the ever-present threat of war with the Soviet Union "created a climate of fear. It stimulated technological progress, provoking an antimodernist backlash . . . by equating scientific progress with national survival. It also encouraged researchers to abuse the rights of students and others. It fed the hubris of some psychologists, encouraging them to seek ways of modifying behavior to make people better citizens. It gave birth to a drug culture and to a generalized disillusionment with America and government."[18]

Kaczynski transformed his frustrations into an ideology, and his experiences at Harvard proved crucial to that ideology and his understanding of the world. It was Murray's experiments that triggered in Kaczynski a suspicion of psychology and the system of which it is a part. Soon nightmares about psychologists followed. Ironically, it was probably Dr. Murray—"an establishment figure who fancied himself a father figure to students—who became a catalyst for transforming Kaczynski's anger at individuals into philosophical fury against industrial society and the central role that psychology plays therein. As the only leading psychologist Kaczynski knew personally, Murray represented the establishment. And this establishment, Kaczynski thought, threatened liberty and demanded submission . . . and retaliation."[19]

Kaczynski would go on to isolate himself from society, living in a remote Montana cabin without electricity and running water, learning survival skills, and beginning a nearly two-decade-long campaign against the modern industrial world. His love of mathematics would be replaced by an interest in explosives, and he would mail sixteen bombs to an assortment of targets including airlines and universities. His one-man terrorism campaign would result in three dead, twenty-three injured, and the haunting specter of the Unabomber's next target.

There are those who believe Kaczynski's long and convoluted political manifesto, *Industrial Society and Its Future*, shows definite indications that his time as a Henry Murray stress subject had an impact on him. Though others in Murray's psychological studies didn't take up arms against their country, Kaczynski's youth and fragile mental state made him too vulnerable for the emotional pounding the professor's experiment was designed to deliver.

**WENDELL JOHNSON AND HENRY MURRAY** were not the only psychologists who initiated experiments that heaped excessive psychological pressure on vulnerable children. A cursory look at psychological journals over the years discloses numerous studies that subjected already burdened children to questionable levels of stress.

In one 1969 study, for example, researchers decided to take an eight-year-old severely retarded boy who had been on tranquilizers, an eight-year-old retarded girl who was forced to wear restraints, and another eleven-year-old retarded boy and use various degrees of aversive therapy to disabuse them of self-destructive and psychotic behavior. Because drugs, ECT, and other treatment regimens proved unsuccessful, doctors decided to experiment with painful shocks delivered to the child's leg or a 1,400-volt shock from a one-foot-long rod with spikes. The authors said the shocks were "definitely painful" and similar to "a dentist drilling a tooth."[20] Later the authors would address the perceived need to ensure that the pain threshold was severe enough to impact children who had already grown used to pain from their own behavior. "To avoid selecting a neutral shock, or a weak one to which the children could adapt quickly," the authors designed one that "smarted like a whip or a dentist drilling on an unanesthetized tooth" to ensure "the subjects gave every sign of fear and apprehension."[21] Incredibly, researchers found it more convenient to use an electrified cattle prod on retarded children than to experiment with less harmful procedures to reduce their self-destructive habits.

Even more recently, in another study twenty-one young males between the ages of eight and eleven who were recruited from schools for special needs children, "stress was induced for 75 minutes in a task that involved frustration, provocation, and aggression."[22] Remarkably similar to Murray's Harvard study, competition between each subject and a "videotaped opponent of similar age and sex" would compete for "best performance over the session. Frustration was induced by having the subject solve a difficult task under time pressure." The task was "made to be unsolvable," especially given the added "provocation" that took the form of constant and disparaging criticism of the subject's performance. Their biological reactions to this stress as measured by cortisol levels and cardiovascular response were compared to that of thirty-one "normal" children.

In another study from the 1990s, children from psychiatric clinics between seven and thirteen years of age—many who were emotionally disturbed or with conduct disorders—were told to imagine being chased by a monster or being scolded for something they had not done, and their biological reactions were also compared to "normal" children.[23] In one 1989 study, eighteen anxious children hospitalized for enuresis or a tonsillectomy were exposed to a supposedly mild stressor while being measured for salivary cortisol and electrodermal activity. The stressor was a series of risky, harrowing scenes from a film called *Rollercoaster*. A control group was shown ducks swimming on a small lake, and differences in biological reactivity to stress between "disturbed" children and normal controls were assessed.[24]

Finally, just a few years ago, twenty-seven children with varying degrees of aggression and anxiety issues were emotionally challenged with a series of go/no go tasks designed to cause emotions such as sadness, anger, and frustration, and their cortical activity as measured by EEG was compared to fourteen normally developing controls.[25] Experimental psychological studies such as these are commonplace in the academic literature, but there are few examples of journal editors, psychological professionals, or readers voicing their concerns about the judiciousness of putting especially vulnerable children in such harmful scientific research. Where are the institutional review boards that were designed to ensure that such abuse did not occur? And where were the journal editorial boards who might have ensured that such problem-ridden experiments were not accepted for publication? Regrettably, the literature is replete with nontherapeutic experiments that could only cause harm.

# TEN

# REPRODUCTION AND SEXUALITY EXPERIMENTS

## *"They Treated Those Girls Just as if They Were Cattle"*

*We did not decide that we would not inform [the women]. We simply felt it was unnecessary.*

—Dr. William Darby

**THOUGH HE HAD DIED DECADES EARLIER, CHARLES DAVEN-**port, the ardent and indefatigable early twentieth-century eugenicist, would have appreciated the gung-ho race-cleansing spirit that dominated certain parts of Alabama in the early 1970s. Nearly four decades long, the US Public Health Service's experiment in Tuskegee monitoring the long-term consequences of "untreated syphilis in the Negro male" was not the only research project in the Heart of Dixie that smacked of eugenics. In the state capital of Montgomery just twenty-five miles away, another government study was under way that in both design and practice fostered all the principles and goals championed by America's better-breeding movement earlier in the century.

In an effort to restrain the birth rate in certain parts of the community, especially among people on welfare and residing in low-income public housing projects, the Montgomery Community Action Committee in conjunction with the federal Office of Economic Opportunity had begun giving experimental birth control injections to women. The drug chosen was depot medroxyprogesterone acetate, which would become better known to the American public as Depo-Provera.

The drug was developed by Upjohn scientists in the mid-1950s. In 1960, Upjohn would apply to the Food and Drug Administration (FDA) for its approval as a treatment for endometriosis. During testing in Brazil, however, it became clear to Upjohn researchers that the drug had utility as a long-term contraceptive as well. That realization led them to apply to the FDA for the drug's approval as a viable contraceptive.[1] During the 1960s, animal and human trials were begun both domestically and in numerous foreign countries. The results disclosed unwanted side effects including tumors and various forms of cancer in dogs and monkeys.[2] One large Upjohn trial in 1967 took place in the Grady Memorial Family Planning Clinic in Atlanta, Georgia, where as many as 1,000 women may have been involved.

Though it had not been blessed by the FDA as an approved contraceptive, more and more doctors were prescribing it off label; they viewed it as a cheaper and more consistent alternative to other forms of contraception. States from Tennessee to Texas were incorporating it in their birth control programs. Poor black women in the South tended to be the overwhelming recipients of the drug. Threats, deception, and the loss of welfare benefits were recurring themes in many of these anti-pregnancy programs.

Anna Burgess, for example, was a twenty-year-old woman from Cumberland County, Tennessee, who lived in a three-room house without running water or electricity. In July 1971, she was called to her local welfare office. "The welfare lady asked me if I had signed up for family planning," recalled Burgess before a congressional hearing in 1973. "I was not on it then, and she said I ought to be taking something or other. She said they'd rather feed one youngun [sic] as two. She made arrangements for me to go to the health departments. She said the shot would last 6 to 8 months. I didn't go the first two times she arranged it because I was scared a little." Burgess went on to tell the Senate Committee, "If it had not been for them, I would not have took it. If it had not been for the welfare people, I would not have taken the stuff in the first place. They wanted me to take the birth control shot so there would not be no more children."[3]

It became increasingly clear to her that if she did not do what they wanted, she would be in serious financial trouble. "Well, it seemed the way I took it was, I thought they would take the check or something. From the impression I got, if I did not take the birth control, they would take the check."

Such stories were commonplace throughout the South in the early 1970s. It was estimated that the state of Tennessee alone had "between 1,000 and 1,500 women getting Depo-Provera each year." And girls younger than twenty were caught in the anti-pregnancy dragnet. As Jessie Bly, an Alabama social worker who was to become a key player in a landmark medical abuse case, said of the widespread, discriminatory, and coercive government practices at the time, "They had become a way of life for many in the South."[4]

One family particularly hard hit by these modern-day eugenic practices was the Relfs of Montgomery, Alabama. Basically unschooled field hands with an assortment of learning disabilities, Lonnie Relf and his wife struggled to raise and care for their three young daughters, all of whom had some degree of mental, physical, and social obstacles to overcome. Katie was the eldest, followed by Minnie and Mary Alice, both of whom were intellectually challenged.

When the family walked through the doors of a Montgomery Department of Human Resources office in desperate need of housing in 1972, Jessie Bly was assigned to assist them. Bly was thirty years old, married to a military serviceman, and had attended Central Texas College before her husband was transferred to a base in Alabama. "Home visits" would become a core part of her social work duties, which included checking on the condition of the elderly and poor and making sure their needs were met. Many would have gone without food, heat, and other necessities if she had not repeatedly gone to their residences to check on them.

Bly helped the Relf family get an apartment at Smiley Court, a public housing complex in Montgomery. While aiding them with the move and doing the paperwork that would assist them in procuring additional government services—not an easy chore considering that the Relfs could neither read nor write—she was particularly struck by the young Relf girls. Katie was just fifteen and already receiving shots of Depo-Provera. Bly suspected that Minnie, a couple years younger, might be on them as well. She was most concerned about the youngest daughter, Mary Alice, who was twelve. She was retarded and missing part of her arm, which had been amputated shortly after birth because the umbilical cord was wrapped around it during delivery. The girls had modest to no schooling, leaving them terribly behind other children their age. Mary Alice, in particular, had a number of issues including "a problem with hygiene" that Bly worked to address.

Lonnie Relf and his wife had to be taught almost everything. Some mundane features of life were mysteries to them. The family had "huge power bills," said Bly, and she couldn't understand what was causing them. An investigation and a couple of visits turned up the reason: "The Relfs had no comprehension of thermostats. They never knew what a thermostat was or how it worked," recalled Bly. "When I went to visit them the heat was always way up, they had it blasting. And all the windows were open to cool the place off. They really had no idea what a thermostat did. They were really backward and completely dependent on public assistance for food, shelter, and medical care."

Life for the Relf family had always been a struggle, but the situation grew particularly dark for the family in 1973. All three girls were receiving Depo-Provera injections by that time, and in March, a nurse took Katie to a health clinic for insertion of an intrauterine device. Her parents were never asked if they had any objection, and Katie certainly didn't want it, but as with most poor blacks in the South, they believed they had to do what the government wanted. In June, a family planning nurse picked up Mrs. Relf and the younger girls and transported them to a doctor's office. "Mrs. Relf was told the girls were being taken for some shots. She thought the shots were the same as those all three children had been receiving for some time."[5] From there they were transported to the hospital, where the girls were assigned a room, and Mrs. Relf was asked to sign a document. Unable to read or write, but believing it had something to do with the girls getting more shots, she placed an "X" on the paper. It was actually an authorization form for "surgical sterilization."

At some point during their brief stay at the hospital, Minnie managed to borrow a dime from another patient on the ward and call a neighbor—the Relfs had no phone—to plead for their mother to come pick them up. Minnie was in fear for her life and believed something dreadful was going to happen to her and Mary Alice. Their mother said she had no means of transportation; she couldn't get to the hospital nor could she get the girls home. According to Joseph Levin, general counsel of the Southern Poverty Law Center, "the next morning both children were placed under a general anesthetic and surgically sterilized. At no point prior to the surgery did any physician discuss with the girls or their parents the nature or consequences of the surgery to which Minnie and Mary Alice were about to be subjected."[6]

That same morning, Jessie Bly arrived at the Relf apartment to take Mary Alice to her first day of school. Bly was intent on getting the youngest Relf girl at least a modicum of formal education and was totally unaware of the visits by the family planning nurse. Katie, the eldest sibling, explained what had taken place. Katie said they wanted to take her as well, but she had locked herself in her room and then ran away so they wouldn't get her. She was scared to death of doctors and hospitals.

Bly went directly to the hospital only to be met by a shattered and terrified young girl. Mary Alice had just come out of surgery and was "scared to death," recalled Bly. "It hurt me so bad. She was so frightened and fearful. She wouldn't let me go. It was pitiful. They treated those girls just as if they were cattle."[7]

Bly was furious. "I went to the doctors, but they wouldn't tell me anything. I then went to the head nurse and demanded to know who authorized the surgery. She told me, 'They could get pregnant. There are boys hanging around them now. We don't want any more of that kind.'"

"That did it," said Bly, who was already greatly disturbed that young girls throughout the state were being injected with experimental birth control drugs. "That's what pushed me out there. It became a matter of principle for me. These people were so backward; they were easy targets. They needed somebody to speak up for them. That's when I decided to talk to a Jesuit priest who did good work in the community. I also got them a good lawyer."

Bly went to the Southern Poverty Law Center, an organization known for helping the poor and dispossessed. When Joseph Levin, the center's general counsel, heard the story, he immediately took the Relfs' case. He would later recount before a Senate subcommittee that the Relf girls had been taken to the clinic because of "new policies which prevented nurses from going into the community to administer shots and birth control devices . . . the only way to insure against pregnancy was sterilization."

At the hearing, Senator Edward Kennedy asked Levin whether he believed they were getting Depo-Provera prior to when they were sterilized. To quote the hearings:

> *Mr. Levin:* I am not familiar with the drug, but I am told that it is the only
> birth control injection available. So I assume that that is the only injec-
> tion they received prior to the sterilization.
>
> *Senator Kennedy:* Did you know that that is an experimental drug, as well?
> Depo-Provera?
>
> *Mr. Levin:* I have since been informed of that.[8]

Levin thought it possible that Congress's effort to deal with one problem—
off-label use of a potentially dangerous drug (Depo-Provera)—may have
resulted in some communities taking the extraordinary step of sterilizing a
large number of poor women. Levin and many others believed young girls in
Alabama were caught in a no-win situation: Their choice was either an experimental drug that they had no understanding of or sterilization.

"I decline to engage in debate over the relative merits of sterilizing children," Levin testified. "I see no justification for permanently depriving any
child of his or her right to conceive, regardless of the child's present mental
or physical condition, nor do I believe that agencies, by committee or other
means, have the right to sterilize any persons, regardless of age, unless that
person, intelligently and with full and complete knowledge of the consequences, desires to be permanently stripped of his ability to create life."[9]

Levin described the lives of many of these poor women who depended on
the state for food stamps, medical assistance, and the "$156 each month from
the Alabama Department of Pensions and Security" for their existence. But
the trade-off was that their lives were now "under a microscope." They were
constantly under supervision with almost weekly visits by some government
representative. As Levin told the lawmakers, "They are surrounded by a welfare state upon which they depend for their very existence, and they are easily
'coerced' into doing what the welfare people recommend to them. It is a very
sophisticated, probably unintentional, form of coercion, but it is extremely
effective."

When questioned on related subjects, such as how widespread such
practices were, Levin expressed doubts as to whether paying customers
would have been treated in such a fashion. "I believe this subcommittee
will find that the sons and daughters of middle America are not sterilized,
regardless of physical or mental condition. It is the 'free clinic' patient who
is fair game for this most final of birth control methods. . . . Sterilization is

not 'birth control' when applied to minors and incompetents—it is mayhem, and it should be stopped now."

It would appear that the early history of Depo-Provera and its misuse and overuse were due not so much to aggressive pharmaceutical company marketing as to average doctors becoming wedded to the notion of the drug as a quick, effective, and relatively cheap method of contraception. Disregarding the fact that it was an investigational new drug still under examination by the FDA, they casually prescribed it for their patients—the majority of whom were poor and black—for a good number of years during the late 1960s and early 1970s.

Even when forced to testify at Senate hearings in 1973, doctors continued to deny Depo-Provera's official status as an investigational drug. As one physician boldly stated, "We keep referring to Depo-Provera as an experimental drug. It has never been our understanding that it is an experimental drug, and our use of Depo-Provera has not been within the context or the framework of the way we would go about doing an experimental study if we did one."

Obviously irritated, Senator Kennedy fired back, "Just to clarify our terms. . . . Depo-Provera is an experimental drug for the purpose of birth control." Earlier in the day, Kennedy pointed out there had been testimony stating that Depo-Provera "is not to be used for birth control purposes."

The combative witness, Dr. James Brown, the superintendent of Arlington Hospital in Arlington, Tennessee, continued to argue the point. "We were advised by our medical specialists that quarterly injections of a progestational product of the Upjohn Company, Depo-Provera, while not specifically licensed for the purposes mentioned above, would safely control menstruation and serve as a contraceptive should sexual exposure occur."[10]

Further questioning by Kennedy concerned its use in state institutions. Brown not only admitted that the drug was used in state facilities for the "retarded" but went on to argue, "The problem of providing birth control, and also the problem of menstrual periods in a facility among the very severely retarded is a considerable problem. It is one we agonize over from the standpoint of the normalization principle, and try to weigh the benefits of obliterating menstruation to the resident, as compared with the benefits to the staff."[11]

Kennedy summed up the situation: "It is my own conclusion that there have to be some very direct and important protections for the individuals that are being affected by these very dramatic and in many instances constructive and others destructive experiments which are taking place."[12]

The case that Joseph Levin, Morris Dees, and the Southern Poverty Law Center brought against the Department of Health, Education, and Welfare (HEW) and the Office of Economic Opportunity would address and ultimately rectify many of these egregious practices that had affected the Relfs and many other families in the South.

In *Relf et al. v. Weinberger et al.*, Judge Gerhard A. Gesell ruled that HEW guidelines had to be corrected, including the need for a definition of the term "voluntary," the lack of safeguards to ensure that sterilizations were indeed voluntary, and the absence of prohibitions against the use of coercion in obtaining consents.[13] On April 18, 1974, HEW published revised regulations that included the changes Gesell had ordered. Informed consent was now defined as "the voluntary, knowing assent" of any person undergoing sterilization procedures verified with a consent form that included information on the actual procedure, any possible risks or discomforts, any benefits of the operation, information on alternative methods of birth control along with an explanation that sterilization is an irreversible procedure, and a statement "that the individual is free to withhold or withdraw his or her consent to the procedure at any time prior to the sterilization without loss of other project or program prejudicing his or her future care and without loss of other project or program benefits to which the patient might otherwise be entitled."[14]

The revised regulations also dictated that every sterilization consent form exhibit prominently at the top of the form this legend: "NOTICE: Your decision at any time not to be sterilized will not result in the withdrawal or withholding of any benefits provided by programs or projects receiving federal funds."

INTERESTINGLY, ONE AGENCY THAT WAS NOT obligated to abide by these regulations was the Indian Health Service (IHS), which performed twenty-three sterilizations of women under the age of twenty-one from midsummer 1973 to April 30, 1974, despite the HEW moratorium on them. Thirteen more sterilizations on underage girls occurred between April 30, 1974, when HEW published the new regulations in the *Federal Register,* and March 30, 1976. It was suggested that some IHS physicians did not completely understand the regulations and that contract physicians were not required to adhere to them.

Health care, in general, for Native Americans is something of a national embarrassment, and women's health concerns were certainly no exception. Sterilization was common on Indian reservations in the 1970s, and in the 1990s the use of Depo-Provera and Norplant (another chemical anti-pregnancy medication) was routine. It is estimated that up to 80 percent of women were "voluntarily" sterilized on some reservations—even women under twenty-one and even within seventy-two hours of giving birth, which violated federal regulations.[15] According to accounts, two fifteen-year-old girls were sterilized during what they were told were tonsillectomy operations. Additional breaches of the regulations included parents not being informed of the procedures, coercion being used to acquire their signatures, improper consent forms, and not observing the seventy-two-hour waiting period between signing and the surgical procedure.

It was also argued that informed consent papers were practically worthless, as those giving their consent could not understand them. And Depo-Provera, still considered dangerous and an investigational drug, was being given to Indian women long after it was no longer being prescribed for women in the rest of the nation.[16]

**AS WAS POINTED OUT IN EARLIER CHAPTERS,** sterilization and castration were once thought viable remedies for troubling sexuality issues concerning "defectives" who had just entered their teenage years. Physicians under the influence of an emerging and boisterous eugenic movement practiced a range of experimental procedures to rein in the explosion of defective germplasm that was gradually laying waste to society. Some in the profession at the turn of the twentieth century were on a crusade to seek out and identify "degenerates" in the community. Extreme measures were occasionally utilized to neutralize their spread.[17] Many converts to the eugenic banner were proud of their work and boasted of their triumphs in various medical journals.

Disturbed by the level of masturbation occurring at his Baldwinville (Massachusetts) home for children, Dr. Everett Flood castrated over two dozen boys between the ages of seven and fifteen. He would go on to claim that the first castrations were performed with the idea of "preventing masturbations in certain cases where the habit was most constant and the boy had no sense of shame, besides being a confirmed epileptic and, of course, somewhat feeble-minded."[18] The editors of the *Journal of the American Medical*

*Association* apparently saw nothing unethical or improper about such drastic and irreversible measures when they published Flood's article. There is no telling the impact of such messages on the profession, but colleagues must have taken notice of Flood's success. "Masturbation has ceased," he informed readers, and the boys were now "more manageable, less inclined to quarrel, and more capable of reasoning."[19]

Flood was a serious champion of the therapeutic effects of castration. He not only articulated its benefits when performed on oxen, horses, sheep, and, of course, humans but also came to the rescue of doctors who might have drawn the wrath of critics for performing such surgery on children. For instance, in one psychology journal article he wrote:

> Dr. Pilcher, Supt. of the Institute for Imbeciles and Weak Minded Children, at Winfield, Kan., has been bitterly denounced by newspapers in Winfield and Topeka for castrating several boys, inmates, who were confirmed masturbators. His predecessor, Dr. Wile, had treated these boys five years without benefit, and Dr. Pilcher, taking a rational view of the subject, performed the operation for the same reason that he would perform any other surgical operation—for its curative effect.

Another doctor who advocated for greater use of such methods was Harry C. Sharp. An ardent eugenicist and surgeon at the Indiana Reformatory, as discussed in chapter 2, Sharp wrote a paper titled "The Severing of the Vas Deferens and Its Relation to the Neuropsychopathic Constitution."[20] He practiced what he preached and cut the vas deferens of forty-two patients whose ages ranged from seventeen to twenty-five to prevent the birth of criminals, cure excessive masturbation, and bring about positive results for the patient. A lobbyist as well as a practitioner of sterilization, Sharp often proposed the mass sterilization of the "feebleminded" and encouraged state authorities to sterilize the 300 girls in the state institution for the "feebleminded."

Those in the medical profession preaching greater utilization of sterilization and castration were preoccupied with other sexual dilemmas besides masturbation, or what was diplomatically referred to as the "sexual over-excitation" of delinquent boys. Those who had bought into the theory of biological determinism viewed castration as a crime preventive. Whereas sterilization served chiefly to prevent offspring, some argued that castration was a crime-prevention measure for the existing generation and indirectly for posterity.[21]

By World War II, the medical profession had, in general, repudiated castration as an acceptable option, but in some quarters it was replaced with hormone therapy, particularly among physicians exploring the treatment of homosexuality for individuals displaying violent or criminal behavior. In one experiment with teenagers at a child guidance clinic, two thirteen-year-old males with effeminate characteristics and one fifteen-year-old with "latent and innate homosexual tendencies" received "androgen hormone treatment." Sounding much like the triumphant Drs. Flood and Pilcher a half century earlier, the author of the article detailing this treatment believed that any individual given testosterone injections "became a pleasing type of aggressive male. There were no more delinquencies. His sexual life also became normal."[22]

**MEDICAL EXPERIMENTS ON PREGNANT WOMEN** had a dramatic impact on the well-being of their fetuses and children. Since the days of Dr. J. Marion Sims, the renowned antebellum surgeon, doctors have found it convenient to experiment with pregnant women. The son of a poor South Carolina family, Sims would graduate from Jefferson Medical College in Philadelphia and return to the South to practice his craft. Sims may have questioned his own competence as a physician early on, but once he commenced his career on the plantations of Alabama, he cast aside all caution. In fact, he would become quite daring. In a terribly misguided attempt to address black infants suffering convulsions, for example, Sims performed radical brain surgery on slave children. He blamed their deaths—or permanent childlike behavior if they survived—on "the sloth and ignorance of their mothers and the black midwives who attended them."[23]

In an attempt to understand and treat vesicovaginal fistula, a catastrophic complication of childbirth, Sims experimented on his own slaves, whose lives were considerably less valuable than those of genteel Southern society. Sims spent four years experimenting on black women with new equipment and new methodologies in the hope of arriving at a surgical breakthrough and making a name for himself in the medical arena. The procedures were terribly painful, and test subjects were given little in the way of anesthetics, but Sims pushed on firmly in the belief "that blacks did not feel pain in the same way as whites."

Publications on the subject by Sims resonated in the medical community, and eventually he would become known as the father of American gynecology.

By the 1850s, Sims had relocated to New York City and become the toast of the town for his well-publicized medical triumphs, although detractors believed the good doctor had done much damage over the years. In trying to determine whether Dr. Sims was a hero or villain, one scholar has written, "It is certainly ironic that an icon of medicine like Sims could be mentioned in the same context as Nazi medical experimenters and the authors of the notorious Tuskegee study on syphilis. An exploration of the apparent paradox reveals as much about the state of medicine during Sims' lifetime as about the man himself."[24]

**ON PREGNANT WOMEN,** the impact of teratogenic agents, such as thalidomide, diethylstilbestrol (DES), and radiation, is especially devastating, resulting in severe consequences for both the women and their children. The thalidomide episode of the early 1960s is one of the most unfortunate in terms of damage done. To those who lived through the period, just the mention of the word "thalidomide" draws the heartbreaking recollections of photographs showing children born with arms like flippers, legs like boat paddles, and severe eye and ear deformities. A drug fiasco and human disaster of incredible proportions, it need not have happened. By the late 1950s, science had evolved enough that it was generally understood that drugs ingested by pregnant women could affect the developing fetuses, even those drugs labeled as "the drug of choice for pregnant women upset by morning sickness."

As the *Sunday Times of London* said of the health calamity:

> It is a popular fallacy to assume the idea that the thalidomide tragedy alerted the world to the danger that drugs can cross the placental barrier and affect the fetus. This was commonplace knowledge before the thalidomide tragedy; and before it drugs were also tested on pregnant animals and given clinical trials on pregnant women. The knowledge and scientific procedures to give protection were available. The disaster might well have been averted everywhere. Unarguably, the havoc wrought could have been much less than that which occurred.[25]

The result of what was referred to as "a huge and disastrous scientific experiment" was untold damage to the German people and people around the world. As one appalled observer stated, "It was an experiment conducted with unusual incompetence, and its dire effects penetrated eventually to most parts of the medically developed world." Seductive advertising copy for the marketing campaign mentioned that the drug was "completely non-poisonous" and "astonishingly safe," and that thalidomide could "be given with complete safety to pregnant women and nursing mothers, without adverse effect on mother or child."[26]

Yet by 1955, if not sooner, it was scientifically quite well known that any substance with a molecular weight of less than 1000 could be expected to cross the placenta and appear in the fetal blood—and the molecular weight of thalidomide is only 258. A general practitioner could not be expected to be aware of this by 1961. But surely Grunenthal (the German pharmaceutical company that produced it) should have been aware of it.[27]

By misrepresentation, the company had made it appear that experimental work had been successfully accomplished. In fact, such work had not been done. But if it had, the teratogenic effects of thalidomide (teratogenic means "monster making," or the creation of major deformities) would almost certainly have come to light between 1956 and 1958. What had become evident were complaints of neuritis when used as a general sedative, but the German manufacturer, Chemie Grünenthal, actively fought those complaints as well as attempts to prevent the drug from being marketed. It squashed early negative news reports, but by 1962 there were rapidly increasing accounts of babies being born with a laundry list of appalling birth defects.

Deformities depended on what stage of development the embryo had reached when the mother took the drug. Stunning physical deformities were the signature feature of a thalidomide baby. As one author mused, "It was as though the jigsaw of life had been jumbled and then the pieces forced into places they could not fit or simply left out altogether."[28]

Fortunately, a young FDA official had significant questions about thalidomide. Even though it was her first drug review assignment, she would not be bullied into granting approval for the drug to be marketed in America. A Canadian pharmacology school graduate from McGill, Dr. Frances Kelsey understood German, was able to read the original application forms, and understood the potential effects of new drugs on developing fetuses. Kelsey rejected the company's applications six times; she had serious reservations about the drug's safety. Unfortunately, the Richardson-Merrell Company, which had gained the rights to market thalidomide in the United States if FDA approval was secured, managed to distribute a large number of pills for investigational purposes—a custom permitted under US law pending agency approval.

As one scholar has written of that loophole:

> The law at the time also permitted pregnant women to be included as research participants after the first three months of the investigational period. Two and a half million tablets were distributed in the U.S. for experimental purposes and nearly 20,000 patients received thalidomide, including several hundred pregnant women. The window of danger for congenital problems turned out to be quite narrow, but the amount needed to cause problems was very small: infants of women who took thalidomide (one tablet was sufficient) between the 20th and 36th day after conception were at risk for malformation. The result, thanks to Kelsey, was that only 17 children in the U.S. were born with thalidomide-related deformities. By contrast, in the rest of the world, an estimated 8,000 to 12,000 infants were born malformed due to their mother's use of thalidomide and . . . only about 5,000 survived beyond childhood. Approximately 40% of thalidomide victims died before their first birthday.[29]

**ANOTHER WELL-KNOWN EXAMPLE** of drugs for pregnant women doing damage to their offspring was the synthetic hormone DES. It was given to women from the 1940s until the 1970s, when physicians finally realized that the drug not only did not significantly prevent miscarriage—the reason it was prescribed—but also substantially increased health risks to the 6 million children born to these mothers.

In utero exposure of the female fetus genital tract to DES often led to malignant transformation.[30] Adenosis, a nonmalignant vaginal problem, occurred in 50 to 90 percent of DES daughters. The relationship between mothers taking DES and their daughters developing rare clear-cell adenocarcinoma of the vagina some fifteen to twenty-two years later is estimated to be between 1 in 1,000 to 1 in 10,000. Other adverse effects for the daughters include infertility, spontaneous abortion, ectopic pregnancy, stillbirths, early menopause, cervical cancer, and breast cancer. Sons of women taking DES are more prone to have undersize sex organs and low sperm counts, epididynal cysts, microcephallus, and testicular hypoplasia, among other health issues.[31]

Use of the drug had not started with such bleak expectations, however. It was once believed that DES averted late pregnancy complications. Many doctors recommended it. As one 1940s study suggested, based on the results of 387 women at Boston Lying-In Hospital, even if women delivered early, DES seemed to protect the fetus against death since the fetuses are unusually premature for their gestational age. Researchers thought this was because DES created a better intrauterine environment for fetuses. And though there was a long discussion at the end of the journal article debating various aspects of the study, not one doctor brought up the possibility that DES might have teratogenic effects on the fetus.[32]

By the early 1970s, though, doctors were reporting a link between vaginal tumors in daughters of mothers who had been prescribed DES. The association with clear-cell adenocarcinoma was the first instance of prenatal carcinogenesis in humans. This finding led to a large number of animal experiments dealing with the mechanisms of transplacental carcinogenesis and the effects of exogenous hormones on developing embryos.[33]

The impact has been significant for the sons and daughters of pregnant women who took DES during the quarter century following World War II. Not long after the connections between DES use in pregnant women and problems with the genital/reproductive health of their children were discovered, headlines such as "A DES Victim Tells of Anger, Awareness, and Despair" and the "University of Chicago to Pay $225,000 to 3 over DES" became more frequent.[34] News of large sums of money changing hands in out-of-court settlements as well as offers of free exams to children of all 1,081 women given DES while maternity patients at the university's hospital followed. Similar stories played out at numerous medical institutions around the country.

But if proper precautions had been implemented early in the drug's development, the situation could have been much different. As one reviewer has written of the DES debacle, "Ultimately . . . it serves as a reminder that though the narrow lens of today might reassure us that an intervention is safe, it is only with the wisdom of time that the full consequences of our actions are revealed."[35]

**ONE OF THE MORE INEXCUSABLE EXAMPLES** of scientists cavalierly incorporating pregnant women in dangerous research and thereby putting their unborn babies at risk at the dawn of the Cold War concerns the radiation studies performed on low-income expectant mothers at Vanderbilt University in Tennessee.

As was eventually revealed, from the end of World War II in 1945 to May 1947, over 800 pregnant women were ushered into Vanderbilt's Prenatal Clinic for a thorough physical examination. The clinic patients, usually poor and not well educated, appreciated any attention by university health personnel. Invariably, the hospital visit would conclude with what was described as a nurturing cocktail. Eileen Welsome graphically captured those ominous moments in her book *The Plutonium Files*. Some women wondered if a "cocktail" was a judicious choice so close to the birth of their baby.

> "What is it?" Helen [one expectant mother] asked.
>
> "It's a little cocktail," said the doctor. "It'll make you feel better."
>
> "Well, I don't know if I ought to be drinking a cocktail," she responded, her voice light and bantering.
>
> "Drink it all," he told her. "Drink it on down."[36]

As it turns out, the young woman who was part of this exchange, and the hundreds of others like her, might have been luckier if they had in fact been handed a shot of bourbon. In actuality, they had been given something far more debilitating and dangerous for both themselves and their fetuses: a glass of liquid containing radioactive iron isotopes. As Welsome states, "[M]any of the women were led to believe that the drinks contained something nutritious that would benefit them and their babies. But nothing could have been farther from the truth. The drinks actually contained varying amounts of radioactive iron. Within an hour the material crossed the placenta and began circulating in the blood of their unborn infants."[37]

The research, coordinated by some of the top biochemists in the country, including Paul Hahn, who was bright, energetic, and described as possessing "insatiable curiosity"—traits not unlike those of many other top-flight microbe hunters—had a definite scientific goal in mind. The scientists were eager to learn how a woman's diet and nutrition affected her pregnancy and delivery and the condition of her infant.

Though some of the more experienced physicians probably understood the implications and potential risk in giving pregnant women radioisotopes, some were just carrying out an assignment, ignorant of the damage they were causing. Dr. William Darby, for example, one of the physicians who doled out the isotope-laced drinks—as one would give "sweet[s]" to a child—admitted years later that the radioactive libations had "no therapeutic purpose" at all. The study was designed to obtain additional knowledge. In fact, he further admitted, he "didn't know much about radiology."

The repercussions of such clandestine and cavalier scientific mischief cannot be overestimated. Manhattan Project physicians may have acquired some useful information, but many of the Vanderbilt women and their children would suffer bizarre health complications that ran the gamut from strange rashes and bruises to unusual blood ailments, including cancer. Welsome's gripping account of one girl developing "a lump about the size of an orange on her upper thigh," whose poisonous contents gradually "spread into the child's spine, then moved up through her lungs, heart, throat, and finally, into her mouth," leading to eventual paralysis and death at eleven years of age, is heart wrenching.[38] Other children of Vanderbilt mothers who developed liver cancer, acute lymphatic leukemia, and synovial sarcoma and died between five and eleven years of age were poignant proof that there was a very likely cause-and-effect link between the covert isotope experiments and the pregnant women who were purposefully deceived in order for them to become raw material for research.

**NO MATTER WHAT WAS BEING TESTED** and for what reason, some doctors found it convenient to experiment on prospective mothers and on children as if they were nothing more than lab animals. Though there was the occasional protest, most of these egregious experiments were met with indifference by both practitioners of the healing arts and medical journal editors. Apparently it was easier to go along with moral and ethical outrages than it was to oppose them.

# ELEVEN

# RESEARCH MISCONDUCT

## *"Science Actually Encourages Deceit"*

*The roots of fraud lie in the barrel, not in the bad apples that occasionally roll into public view.*

—William Broad and Nicholas Wade

IN 1998, BRITISH PARENTS OF INFANTS AND YOUNG CHIL-
dren were given a severe fright. A London researcher was making headlines
with his claim that the combination measles-mumps-rubella vaccine, known
as MMR, was actually quite dangerous and was precipitating the spike in
autism around the globe. Panicked parents immediately began to question
the wisdom of their family physician, and many decided to forgo MMR shots
for their children. These difficult familial decisions would result in numerous
repercussions, not the least of which were growing doubts about the medical
profession and an ever-increasing number of children falling victim to measles, mumps, and rubella (or German measles).

Andrew Wakefield, the physician who had sparked the controversy regarding vaccine safety and its relationship to autism, had published a study
earlier in the decade linking measles to Crohn's disease but later admitted
he had been wrong. He was now certain that MMR was linked to autism
and recommended that children not be vaccinated. After his report made

worldwide headlines, there was a noticeable decline in the numbers of children receiving the MMR vaccine. The increasing numbers of laypeople supporting Wakefield's allegations, growing consternation among physicians, and general controversy surrounding vaccine safety led to much angst, debate, and, in due time, public hearings on the issue.

The son of a surgeon and general practitioner, Wakefield earned his degree from St. Mary's Hospital Medical School and became a fellow of the Royal College of Surgeons. His specialty would become tissue rejection, especially with regard to small-intestine transplantation. He would eventually become the head of experimental gastroenterology at the Royal Free Hospital School of Medicine.

Wakefield's name would become known to his medical colleagues around the world through his controversial 1998 *Lancet* article. Along with a dozen coauthors, Wakefield's study of twelve autistic children claimed there was a link between bowel disease, autism, and the MMR vaccine. "Autistic enterocolitis syndrome" was the name he gave his new discovery, and he was not shy about broadcasting it. A well-attended press conference and numerous media interviews ensured that his theory received maximum exposure.

Parents of autistic children were desperate for answers, and Wakefield's theory provided an explanation. Curious about his findings, doctors and medical researchers made repeated attempts to replicate a definite connection between vaccines and autism but were unable to do so. After they could not establish the critical linkage Wakefield had supposedly discovered, doubts about his hypothesis began to grow.

More and more parents fearing the specter of autism were deciding to delay normal vaccinations for their children, but grave concerns about Wakefield's research were increasing in the medical community. The doctor's *Lancet* article and MMR theory took a dramatic hit six years later when Brian Deer, an investigative reporter for the *Sunday Times of London*, wrote a scathing article regarding Wakefield's research practices and financial motivations. Deer raised questions about Wakefield's errors. He interviewed some of the parents of the six- to nine-year-old children in the study and found that their stories didn't match the doctor's claims. No case was free of misreporting or alteration. Deer was also critical of Wakefield's practice of putting children under general anesthesia, doing spinal taps, threading fiber optic scopes into their intestines, doing biopsies, and collecting large quantities of blood for testing.

And all this was without Ethics Committee approval.[1] Deer communicated his concerns to the *Lancet's* editor in chief.

More explosive yet, Deer charged that Wakefield had received $100,000 (the actual sum was much higher) from a personal injury lawyer representing a parents' group that believed vaccines caused many illnesses. They were looking for a respected medical advocate to join their cause, and the attorney found one in Wakefield. With this financial conflict of interest now revealed, a number of the original article's coauthors requested that their names be removed from the research. The British General Medical Council was brought in and determined that misconduct had taken place, that autistic children had been subjected to unnecessary medical procedures, and that Wakefield had never received institutional review board approval for his research.

It was not long afterward that a retraction appeared in the *Lancet* asserting that the data were insufficient to support the claim of a causal relationship between MMR vaccines and autism and that no studies were able to replicate Wakefield's hypothesis. The UK General Medical Council examined the charges against Wakefield and concluded that he had acted against the interests of his patients and was irresponsible in his research. In May 2010, he was struck off the medical register, which ended his career as a doctor in the United Kingdom. As one reviewer has commented on the controversial case, "Good science will be reproduced by other investigators; bad science won't."[2]

Though Andrew Wakefield's research and reputation would be seriously damaged—the *British Medical Journal* labeled his work an elaborate fraud—both Great Britain and America witnessed a "vaccine scare" and a steep decline in MMR vaccinations. The result would be a sharp increase in measles among children and unnecessary fatalities. In 2008, for the first time in fourteen years, measles was declared endemic in England and Wales.[3] The *British Medical Journal* estimated that Wakefield's mischief has resulted in hundreds of thousands of children in the United Kingdom being unprotected and vulnerable.

**REGRETTABLY, RESEARCH MISCONDUCT** involving children extends far beyond Andrew Wakefield's shenanigans in the MMR/autism controversy. There are numerous examples of physicians and academics purposefully manipulating data in order to accomplish some grand scheme or goal. Such unethical practices, however, fly in the face of professional principles like fidelity and integrity and fail to consider the damage done to patients and the medical profession. The manipulation of data and the publication of false results waste precious research funds, take up journal space, and provide fictitious and misleading information. The theoretical and scientific detours are bad enough, but some may result in proven prevention strategies being postponed or completely jettisoned, leaving populations at great risk.

Many a scientific investigator has contemplated the motivation to commit fraud. Some believe it is the all-consuming desire to be first with a new discovery. As William Broad and Nicholas Wade have argued in their book *Betrayers of the Truth,* "There are no rewards for being second."[4] They go on to state that, in trying to be first with a discovery, "some researchers . . . sometimes play fast and loose with the facts in order to make a theory look more compelling than it really is. The desire to win credit, to gain the respect of one's peers, is a powerful motive for almost all scientists." As Broad and Wade point out, "Massaging data in some way may help toward getting an article published, making a name for oneself, being asked to join a journal's editorial board, securing the next government grant, or winning a prestigious prize."[5]

They go on to argue there is an almost total absence of credible deterrents. Most cases go unreported or are kept quiet. And on those rare occasions when someone is willing to report a violation or breach of ethics, the accuser is often penalized for being vigilant. For each major fraud, they suggest, a thousand minor fakeries are perpetuated. The reward system and career structure of contemporary science are among the factors that induce fraud. Those impressive rewards overmatch the rather modest chance of getting caught, especially for those who have long lusted after professional success and personal recognition.

The impact of Paul de Kruif's *Microbe Hunters* in the 1920s cannot be overestimated in its power to fire the imagination of generations of aspiring doctors and medical researchers. His heroic account of doctors laboring for years in the laboratory, trekking through godforsaken jungles, and battling everything from deadly mosquitoes and outdated practices to stuffy bureaucrats and lack of financial support, enhanced the reputation of a formerly staid profession and sparked popular interest in the lives and personal crusades of the great disease fighters. There is no telling how great an influence the personal sagas of Koch, Pasteur, and Reed had on the scientists detailed in this book.

Some aspiring scientists, for whatever reason, perpetuate scientific sins. Some misrepresent data; others practice deception, thereby placing subjects at risk. Still others may just remain silent while they observe colleagues conducting an experiment that places an individual in jeopardy, embellishing a journal article, or falsifying test results.

That such behavior is so pervasive and widely accepted speaks to the power of the medical profession in getting its members to rationalize and buy into the system. As one observer has commented, "When a person sees ego-inflation, monetary gain, power, or prestige as the criterion for success, engaging in fraudulent behavior will cause minimal dissonance arousal . . . an individual may proceed to reduce the dissonance through a type of cognitive reframing or rationalization (What will it hurt? . . . those would have been the results anyway)."[6]

**WE COULD CITE NUMEROUS OTHER EXAMPLES** of research misconduct to underscore our point regarding the prevalence and acceptance of such unethical scholarship. Here we'll focus on one of the better-known examples of research misconduct, that of Sir Cyril Burt, an eminent, award-winning British psychologist whose research eventually came under considerable suspicion and was finally rejected.

Between the start of World War I and the onset of the Great Depression—a period of time that corresponds with the height of the eugenics movement—Burt collected an impressive amount of data to support his hypothesis that intelligence was determined by heredity. Though there may have been those who doubted the wisdom of Burt's pronouncements over the decades, it was not until the 1970s and Burt's death that detractors began to actually challenge his scholarship. Leon Kamin of Princeton University was one of the first to criticize missing details and false statements. "Kamin came to the conclusion that Burt has 'cooked' his data in order to arrive at the conclusion he wanted," wrote Alexander Kohn in *False Prophets*.[7] Additional critics would soon surface, leading to increased scrutiny of Burt's work and his formerly unchallenged findings. As more scholars took notice of the inconsistencies, it became all the more amazing that "during Burt's lifetime his work was never challenged despite its shortcomings."[8]

This phenomenon, according to Broad and Wade, has much to do with scientists' becoming prisoners of their own dogma. An entire coterie of scientists appeared ready and willing to accept and digest what was served up to them. The scientific community has too often allowed charlatans and propagandists to impersonate scientists and wave a banner of suspect and illegitimate claims making for bad science and potential harm.

Burt was a disciple of the great eugenicists like Charles Benedict Davenport and Henry H. Goddard. Never wavering in his hereditarian beliefs, he was quite willing to massage his data to confirm his convictions. Early twentieth-century IQ testers, argue Broad and Wade, "had a hereditarian bias so strong that it blinded them to the evidence of environmental influence that cried out from their data. All they could see was the reflection of their own dogmatic beliefs which, just like those of Samuel Morton, echoes the prejudices of their time and social class."[9] The result was multifold, impacting everything from reshaping national immigration policy to changes in the military and other American institutions. Data repeatedly pointed to the importance of environmental factors, but those administering the tests created tortuous rationalizations to preserve their hereditarian biases.

With the passage of years—and still intransigent on the subject of the genetic origin of intelligence—Burt rallied both data and coworkers from his vast reservoir of prejudiced research to undergird his scientific argument and fool his fellow scientists for more than a generation. Even once-loyal Burt defenders eventually realized that he had systematically arranged and rearranged data and studies to document the heritability of intelligence. "This deception is inexcusable for a scientist," admitted Arthur Jensen, but that realization had come decades late, for the ruse had already affected educational policy throughout the nation.[10]

**EMBARRASSING INCIDENTS OF RESEARCH MISCONDUCT** affecting children occurred in fields other than virology and genetics. Charles Glueck, for example, was a well-respected and much-published physician who was in charge of the lipid unit and the General Clinical Research Center at the University of Cincinnati. He was one of the most influential and well-funded scientists at the Center; Glueck published approximately 400 articles, an average of a remarkable seventeen per year after his graduation from medical school.

In 1987, however, the National Institutes of Health (NIH) found that an article in the journal *Pediatrics* was sadly lacking and called it "utterly shoddy science." The paper tackled a controversial treatment for children at risk for developing heart disease.[11] Doctors could not agree on whether a low-fat diet combined with cholesterol-lowering drugs called resins would hinder a child's physical development, causing concern that their growth could be stunted. Glueck, however, declared the diet perfectly safe.

Prior to publication, two anonymous calls were placed to NIH warning of potential problems, but those warnings were never passed on to the journal editor. Though the article was published, a continuing investigation disclosed that Glueck's methods were unacceptable by any scientific standard. He had not measured the children's height, weight, and cholesterol levels, as required for such a study. Eventually the article was retracted, but there is no record of who, under its influence, might have tried to replicate its findings, thereby placing children at increased risk. Glueck was barred from federal funding for two years and resigned his position at the university.

In his defense, Glueck claimed that he was overworked. "When career pressures," he said, "get too great in any field people do funny things. They work hours that would put a securities lawyer to shame."[12]

An argument can be made that Glueck's errors pale in comparison to what some consider the country's six-decade-long experiment with fluoridation, an experiment that violates the Nuremberg Code since people have little choice but to drink the treated water whether they want to or not. As the argument is presented, when the US Public Health Service endorsed the widespread addition of fluoride to the nation's water systems, no trials of its impact had been completed, the sugar lobby was a strong presence, and since the Manhattan Project required huge quantities of fluoride, it was necessary "to change the image of fluoride from a nasty pollutant to something so harmless that children could drink it." But there were allegations that childhood overexposure was culminating in bone fractures, lowered IQ, and lowered thyroid functioning.[13]

In 1950, opponents of fluoridation were few in number, and their questions regarding safety were disregarded, but there is considerable evidence today that fluoridation contributes to fluorosis, or white specks or pitting of children's teeth. Many countries never bought into the fluoridation craze yet their rates of cavities in children are no different from the rates from fluoridated countries.[14] Plans to lower fluoride levels in the natural water supply are under way in many areas due to fluorosis and the increasing evidence that our children are exposed to too much fluoride.

**THE FIELD OF APPLIED PSYCHOLOGY** has not been spared claims of research misconduct. One of the better-known cases is that of Stephen E. Breuning, a University of Pittsburgh psychologist who developed a name for himself as an expert on the effects of behavior-controlling drugs on severely retarded institutionalized persons. Although Breuning never received a doctorate, he achieved a national reputation owing to his many studies and journal publications over the years.

Earlier in his career, Breuning had worked under Dr. Robert Sprague at the University of Illinois. Their research dealt with the effect of neuroleptics (tranquilizers) on violent intellectually challenged patients. After Breuning departed to take a position at the University of Pittsburgh, Sprague began to notice errors if not outright lies in his former apprentice's studies, particularly in Breuning's new research, which argued that withdrawing neuroleptics would increase IQ. The numbers were too good for Sprague's taste, and he feared "there was fakery."[15] Sprague notified the National Institute of Mental Health (NIMH) and the University of Pittsburgh of his suspicions. Though the investigation moved slowly, Breuning eventually admitted faking data under questioning and resigned. At the conclusion of the investigation, NIMH found that Breuning "knowingly, willfully, and repeatedly engaged in misleading and deceptive practices" and banned him from receiving research funds for ten years. As one observer has commented, in cases like this, the work isn't just bad, "it's potentially deadly . . . he's playing with lives."[16] Between 1980 and 1983, Breuning published twenty-four papers on neuroleptics and related topics—a full third of the literature on the subject. Coauthors claim never to have seen his raw data.

The federal court case that resulted in Breuning's pleading guilty forced the University of Pittsburgh to reimburse NIMH $163,000 in grant monies and repay over $11,000 in salary. Breuning was sentenced to serve 250 hours of community service and spend sixty days in a halfway house. One last stipulation was that Breuning cease any involvement in psychological research for at least five years.[17]

The results of the Breuning affair cannot be overestimated. According to Alan Poling, "the recognition that Breuing's work cannot be trusted has seriously eroded the data base concerning psychotropic drug effects in mentally retarded people. We now know less about how psychotropic medications affect this population than we appeared to know when Breuning's data was accepted. This has implications for patients, as well as scientists."[18]

It has been argued that the greatest purveyors of fraud are not young, struggling researchers but experienced, well-published ones, and the problem will only grow worse as pressure mounts to publish groundbreaking articles that foster acclaim, status, and professional recognition. It has become increasingly apparent that science actually fosters deceit in order to build a record of important publications that will earn an individual personal recognition, reap honors and awards, and invitations to join celebrated organizations and associations. Deception, it is now thought by many, will help secure such goals.

# CONCLUSION

**A JOURNAL ARTICLE DURING THE EARLY YEARS OF THE COLD**
War recounted how a number of toxic substances were introduced into the systems of "mice, guinea pigs, and humans."[1] The humans subjected to the toxic substances, however, were not members of the Boston Philharmonic Orchestra, the Massachusetts Chamber of Commerce, or the Harvard University faculty, but mentally challenged young males at the Wrentham State School. Injected with dangerous substances and then routinely bled to observe the negative impact, the Wrentham subjects were labeled "volunteers"—like the mice and guinea pigs.

A few children at the Fernald State School, Wrentham, and other institutions were old enough and had sufficient cognitive ability to understand what was happening, but many others were either too impaired or too young—many were actually infants—to fully comprehend how they were being manipulated. A few, like the State Boys of Fernald, would learn the truth decades after their involvement as test subjects. Most have never learned and probably never will learn how they were once exploited for the benefit of scientific advancement.

Researchers—many motivated by the noblest of causes, others by the prospect of fame and fortune—gravitated to orphanages, hospitals, and institutions for the "feebleminded" when in need of test subjects to conduct a clinical trial. The ethical constraints for such dubious acts were sorely lacking—a direct result of an exploitive ethos that reeked of both eugenics and paternalism. That raw utilitarian spirit, combined with a ramped-up sense of urgency during World War II and the Cold War that followed, contributed to a no-holds-barred atmosphere that fostered scientific research and some valuable discoveries. But those triumphs can't be severed from the tragedy of manipulating and jeopardizing the health of society's least fortunate members.

Physicians tried to rationalize their conduct by arguing the need for isolation, control, and predictability in their research protocols, but they never adequately explained why their choices were always facilities like the Skillman Center for Epileptics, the Iowa Sailors Home for Orphans, and the Wrentham State School for the Feebleminded, over the many prep schools and colleges that held individuals of similar ages and were usually more convenient as a travel destination. Though most would no doubt deny any eugenic influence, the preponderance of evidence suggests that doctors knew which people they could risk harming and with whom they could not take such liberties.

Doctors were either ignorant of or all too willing to abandon medical codes of conduct such as the Hippocratic Oath, American Medical Association directives, and the Nuremberg Code. They routinely trafficked in children to operationalize some bit of medical research. And institutions like Letchworth Village, Vineland State Colony for the Feebleminded, the Pennhurst School, Sonoma State, and St. Vincent's Orphanage were happy to accommodate the great men of science by opening their doors in order to expand our knowledge of disease and improve the human condition. For Mark Dal Molin, Gordon Shattuck, and thousands of other children, that open-door policy would prove the final crushing blow to already grim, compromised lives in austere custodial institutions.

The most zealous of the medical profession were not inherently evil men and women or crass charlatans, but they passionately pursued a particular vaccine or held an intractable stance in favor of one treatment regimen or another that placed children in great jeopardy. Radiation experiments or so-called tracer studies during the Cold War generally incorporated minimal amounts of radioactive substances. Constantine Maletskos of MIT underscored his attempt to use the "smallest amount possible" of radioactivity on the Fernald subjects, but as experts freely admit, not even the smallest dose of radioactivity can truly be called safe.[2] Thyroid uptake studies, however, especially in infants just a few hours or days old, are another story. There is no telling how many individuals died or suffered from rare cancers as a result of those injudicious experiments.

Acclaimed physicians like Lauretta Bender and Walter Freeman were equally blind to the damage they were doing. Bender was too quick to see childhood schizophrenia or autism in troubled youngsters and only slightly less quick to prescribe ECT and LSD as the proper remedy. Freeman was equally determined to cure the world of myriad psychological maladies through the use of crude prefrontal and transorbital lobotomies that he believed didn't even require hospitalization. In the time it takes to boil an egg, he believed he could cure phobias, depression, anxiety, and a propensity for violence. Even children who argued with their stepmother could be helped with a few jabs of a simple ice pick-like instrument. Others like Wendell Johnson had their own pet theories and knew whom to experiment on to establish support for those theories.

This less well-known and occasionally sordid history of medicine's wholesale commodification and abuse of children in human research reminds us of the comment of Yale law professor Jay Katz when assessing the damage inflicted on American citizens in secret Cold War experiments. He wrote that "all this is a frightening example of how thoughtlessly human beings, including physicians, can treat human beings for noble purposes."[3] "Aggression," Katz believed, was "inherent in all of us," doctors and nurses included. It should also remind us that a too ruthless search for knowledge is accompanied by significant human and societal costs, and that children—society's most vulnerable and defenseless group for purposes of research—were often sacrificed in that quest for advancement.

**WE RECENTLY MARKED THE FIFTIETH ANNIVERSARY** of the infamous Jewish Chronic Disease Hospital case in which Sloan-Kettering cancer researcher Chester Southam was allowed to inject live cancer cells into senile and infirm patients. Three conscientious physicians blew the whistle on the project, causing it to be abruptly terminated. We might ask why there have been so few instances of doctors taking principled stands and thwarting unethical experimentation.[4]

Why, one is forced to ponder, were there no doctors during the four-decade-long Tuskegee syphilis study who were willing to expose the fact that sick and dying men were going untreated? Why was no opposition heard after the experiment had been repeatedly written about in medical journals over the years? Why did no one raise concerns about Lauretta Bender's published reports detailing her use of ECT and LSD with young children? Why did the medical profession allow Walter Freeman to travel around the country like a surgical Pied Piper performing lobotomies on adults and children as if he were conducting some routine and innocuous health survey? And why did Wendell Johnson's Iowa orphans have to endure psychically damaging speech sessions without someone standing up and attempting to put a stop to it?

Was there no one at Bellevue Hospital in the early 1940s who recognized that six-year-old Ted Chabasinski was not suffering from schizophrenia and was too young for a battery of twenty electroshock treatments? Was there no one at the Fernald School who recognized that twelve-year-old Charlie Dyer had had enough of the Science Club and was willing to risk serious injury and possibly death by climbing up to the building's rafters in order to avoid any more time as an experimental guinea pig?

Much is written about the "blue wall of silence" in police departments and the inner-city "street codes" that silence witnesses of violent crime, but an argument can be made that the moral amnesia and ethical paralysis of those in the medical and psychological research professions were equally formidable. As Dr. A. Bernard Ackerman often argued, "A conspiracy of silence emerged to protect the profession from the perversion of principles that had taken place in the medical community."[5]

All too infrequently have individuals in the medical community come forward to declare some piece of research mischief unacceptable. As one disenchanted observer commented in 1921 after learning that a New York pediatrician at the Hebrew Infant Asylum had experimentally induced rickets in young children, "No devotion to science, no thought of the greater good to the greater number, can for an instant justify the experimenting on helpless infants, children pathetically abandoned by fate and entrusted to the community for their safeguarding. Voluntary consent by adults should, of course, be the sine qua non of scientific experimentation."[6]

Of course, the research physician had his own thoughts on the matter, and they were based on a utilitarian calculus of convenience, self-interest, and the chances of a grand scientific payoff. The use of the unworthy as test "material" only lessened the odds of getting caught if something untoward should occur. "Research on institutionalized children," wrote one investigator, "offered scientific advantages because the standardized conditions in the asylum approximated those conditions which are insisted on in considering the course of infection among laboratory animals but which can rarely be controlled in a study of infection in man."[7]

One might naturally ask what was different about those few physicians for whom the abuse of patients and test subjects was so offensive as compared to the many others who followed orders or went about their business as if nothing untoward were occurring. Why were there so few who had the moral capacity to discern a wrong being committed and the personal courage to speak up about it? Thousands have suffered as a consequence.

Integrity is just one of the principles that undergird our emerging codes of ethics over the years. In addition to ensuring that researchers do not fudge data, fabricate results, or plagiarize, integrity also means bringing attention—blowing the whistle—to violations of ethical principles. As this book has shown, the principles have been embedded into our research culture with varying degrees of success.[8] Textbooks and college ethics courses may stipulate that subjects have the right to refuse to be incorporated in research and to be free from coercion (autonomy), to go unharmed (nonmaleficence), to have their well-being promoted (beneficence), to be told the truth about an experiment (veracity), to be treated fairly (justice), and to have promises honored (fidelity), but all too often one or more of these principles were jettisoned to the experiment under consideration.

**FOR MOST OF THE LAST CENTURY,** codes of ethics regarding human experimentation were few in number, little known, and rarely enforced. The Oath of Hippocrates and the American Medical Association's first code of ethics, formalized in 1903, were generally ignored or not seen as applicable to medical research. Granted, there were occasionally the sobering pronouncements of the great men of science about research parameters, but the practical impact of these high-minded statements was negligible. Dr. William Osler, one of the founding professors of Johns Hopkins University, may have articulated, "We have no right to use patients entrusted to our care for the purpose of experimentation unless direct benefit to the individual is likely to follow," but how many hospital ward physicians actually practiced this philosophy?

Even Osler, however, understood the limits of high-minded rhetoric, especially when matched against the attraction of scientific experimentation and all that it represented. "Enthusiasm for science has, in a few instances, led to regrettable transgressions of the rules," admitted Osler in 1907, "but these are mere specks which in no way blur the brightness of the picture—one of the brightest in the history of human effort—which portrays the incalculable benefits to man from the introduction of experimentation into the art of medicine."[9]

These "specks"—or embarrassing episodes—that Osler refers to would increase in number and severity, but not so much that the medical profession clamped down and made a concerted effort to rectify the problem. Antivivisectionists would protest and mobilize against researchers and their "specks" of indiscretion, but they were easily marginalized and routinely cast aside as uninformed, highly emotional do-gooders. No legislation that was proposed to protect human subjects during this period came to fruition.

Year after year and decade after decade, medical researchers went about their business with few legal or ethical restraints to harness their insatiable appetite for combatting a dreaded disease or the scientific exploration of some medical curiosity. The confluence of the popularization of scientific heroes—gods of science—and the devaluation of certain institutionalized populations by a clamorous eugenics movement contributed to a potent brew that celebrated triumphant outcomes and dismissed those who were sacrificed in the effort as little more than throwaway people. Few, if any, would contemplate why we so clearly saw a multitude of vices occurring in Nazi Germany's medical establishment during the 1940s but turned a blind eye to our own sins in the use of vulnerable populations as test subjects. Doctors' authority, dedication, and zeal—not to mention their conquering diseases like polio and discovering wonder drugs like penicillin—allowed them a wide berth in scientific endeavors.

Cracks in the morally fragile façade would appear in the 1960s and grow more visible with the revelations of unethical experimentation by Henry Beecher and Maurice Pappworth, the federal government's early effort to establish the concept of Institutional Review Boards, and the various political and social protest movements that were occurring at the time. And finally, the stunning 1972 revelation regarding the Tuskegee syphilis study was the shot across the bow of America's casual indifference to the harm that was being perpetrated on thousands of devalued citizens; a healthy percentage of them we can now say were the neglected and abandoned children housed in stark institutionalized settings.

**THE LAST QUARTER OF THE TWENTIETH CENTURY** would prove an overdue corrective of the laissez-faire atmosphere that had long characterized medical research in America. Experimentation would now be subjected to institutional oversight committees, increased paperwork, assurances that test subjects understood what they were participating in, and additional safeguards against coercion, bribes, and misleading information. Abuse of test subjects in medical research would still take place, but the levels of exploitation and mistreatment were significantly reduced, and both doctors and institutions were now fully cognizant of restrictive codes and regulatory guidelines that established ethical boundaries within which they could work.

One of the most dramatic changes to occur impacting vulnerable populations would be the rather sudden end to the long tradition of using asylums, orphanages, hospitals, and prisons as research mills. With access to cheap and available domestic "material" closed off, investigators and their sponsors began looking abroad for new "fertile fields" of experimental opportunity.

Those far-flung fields of experimental endeavor beyond our shores have proved exceedingly fruitful. With federal regulations inapplicable and far less monitoring of overseas operations, it is currently estimated that 80 percent of all drug approvals are based in part on research data accumulated outside the United States. Institutions like Sonoma State Hospital, the Ohio Soldiers' and Sailors' Orphans Home, the Pennhurst School, and the Vineland State Colony for the Feebleminded have been replaced by China, India, Tunisia, and Nigeria as sites for Phase I drug studies. Drug firms now travel to "places where regulation is virtually nonexistent, the FDA doesn't reach, and the mistakes can end up in pauper's graves."[10]

In fact, the motivating factors pushing Big Pharma and research organizations abroad are practically identical to those that encouraged them to enter America's great human warehouses during the last century. First and foremost is the continuing need for human subjects. As one observer commented, "currently pharmaceutical sponsors are engaged in a turf war over human subjects." The upshot of the offshore research boom is that it's cheaper to do research abroad, where it is easier to recruit test subjects—many of them incorrectly believing they are being treated—and where there is less likelihood that negatively impacted individuals will seek legal counsel.[11]

The gains from these new fields of opportunity have not come without some setbacks, including numerous injuries and death. It should be no surprise that these deaths have had as great an impact on the American consumer as did the medical experiments on the feebleminded in the 1940s and 1950s.

**GRANTED, THERE HAS BEEN MUCH IMPROVEMENT** in the protection of experimental test subjects in recent decades, but any self-congratulatory celebration should be tempered with the understanding that we have not only moved the majority of our drug testing abroad, but we have quite possibly adopted the practices of an earlier era in which society's outcasts were routinely exploited and were only of value as experimental "material."

If we have learned anything from our sad history of using institutionalized children as research subjects, it is our propensity to marginalize our least valued members, our cavalier attitude toward protecting those most in need, and our willingness to jettison inconvenient ethical constraints in order to follow the path offering the greatest rewards. That history, as uncomfortable as it may be, must be acknowledged and made available to forthcoming generations. Moreover, we as a society must be ever vigilant that such paternalistic and utilitarian practices never return, nor be outsourced abroad so that the burden of scientific advancement is relegated to vulnerable populations in lesser developed nations. Scientific progress and the medical advances it fosters is a process we can all celebrate, but the attainment of such triumphs on the backs of children and other powerless groups makes their realization all the less impressive and praiseworthy.

# NOTES

## INTRODUCTION  "THEY'D COME FOR YOU AT NIGHT"

1. Numerous interviews were conducted with Charlie Dyer, both in person and over the phone between 2009 and 2012, about his recollections of the Fernald Training School and his participation in the Science Club medical experiments.
2. Edwin Black, *War against the Weak: Eugenics and America's Campaign to Create a Master Race* (New York: Four Walls Eight Windows, 2003), p. 55.
3. Elof Axel Carlson, *Times of Triumph, Times of Doubt: Science and the Battle for Public Trust* (New York: Cold Spring Harbor Press, 2006), p. 50.
4. James W. Trent Jr., *Inventing the Feeble Mind: A History of Mental Retardation in the United States* (Berkeley: University of California Press, 1994), p. 136.
5. Ibid., p. 163.
6. In addition to the wide net that had been used to capture the many troubled souls residing in asylum and hospital wards throughout the country, specific remedial solutions such a sterilization were still being utilized by eugenics boards around the nation. In North Carolina, for example, the state eugenics board functioned until 1977 in the hope that it would "keep welfare rolls small, stop poverty, and improve the gene pool." Kim Severson, "Thousands Sterilized, a State Weighs Restitution," *New York Times,* December 10, 2011.
7. Soviet premier Nikita Khrushchev's pointed comment, "We will bury you," was not so much a threat of war or nuclear annihilation between the United States and the Soviet Union as a statement of the inevitability of Marxist thought and economics. Most Americans, however, interpreted it as a not-so-veiled threat of impending conflict between the world's two superpowers.
8. Adam B. Ulam, *The Communists: The Story of Power and Lost Illusions 1948–1991* (New York: Scribner's, 1992), p. 52.
9. Dyer interviews.
10. One of the best examples of this phenomenon was Dr. Austin R. Stough, an Oklahoma physician who began a small-town medical practice during the Great Depression. By the end of his career, he had made millions of dollars conducting clinical trials for the pharmaceutical industry in three different state prison systems. Walter Rugerber, "Prison Drug and Plasma Projects Leave Fatal Trial," *New York Times,* July 29, 1969.
11. Quoted in Allen M. Hornblum, *Acres of Skin: Human Experiments at Holmesburg Prison* (New York: Routledge, 1998), p. 37.
12. James Jones, *Bad Blood: The Tuskegee Syphilis Experiment* (New York: Free Press, 1981).

## 1  THE AGE OF HEROIC MEDICINE

1. Paul Starr, *The Social Transformation of American Medicine* (New York: Basic Books, 1982), p. 81.

2. Ibid., p. 143.

3. Paul de Kruif, *The Sweeping Wind* (New York: Harcourt, Brace and World, 1962), 19.

4. Ibid., p. 58.

5. Ibid., p. 59.

6. Lewis rejected the Pulitzer award, calling all such prizes "dangerous." The Pulitzer, according to Lewis, was particularly problematic as it was given to the "American novel that best presents the wholesome atmosphere of American life and highest standard of American manners and manhood." In other words, the prize was "not based on actual literary merit, but obedience to a code of good form popular at the time." *Arrowsmith*—the film—opened in 1931 starring Ronald Coleman and Helen Hayes. The John Ford production was nominated for four Academy Awards and underscored the increasing reverence for medical research. As Martin Arrowsmith tells Dr. Gottlieb at their initial meeting, "I don't want to be a physician, I want to be a researcher." Charles E. Rosenberg, *No Other Gods* (Baltimore: Johns Hopkins University Press, 1997), 123.

7. Paul de Kruif, *The Sweeping Wind*, 115.

8. Ibid., p. 118.

9. Paul de Kruif, *Microbe Hunters* (New York: Harcourt Brace, 1926), 309.

10. Ibid., p. 315.

11. Ibid., p. 349.

12. Ibid., p. 303.

13. Ibid., p. 314.

14. Ibid.

15. Ibid., p. 318.

16. Ibid., p. 315.

17. Ibid., p. 349.

18. Bert Hanson, "Medical History for the Masses," *Bulletin of the History of Medicine* 78, No. 1 (Spring 2004): 150.

19. Gerald L. Geison, "Pasteur's Work on Rabies: Reexamining the Ethical Issues," *Hastings Center Report* 8 (1978): 26–33.

20. Henry Heiman, "A Clinical and Bacteriological Study of the Gonococcus Neisser in the Male Urethra and in the Vulvovaginal Tract of Children," *Journal of Cutaneous and Genito-Urinary Diseases* 13 (1895): 384–387.

21. George M. Sternberg and Walter Reed, "Report on Immunity against Vaccination Conferred upon the Monkey by the Use of the Serum of the Vaccinated Calf and Monkey," *Transactions of the Association of American Physicians* 10 (1895): 57–59.

22. Godfrey R. Pisek and Leon Theodore LeWald, "The Further Study of the Anatomy and Physiology of the Infant Stomach Based on Serial Roentgenograms," *American Journal of Diseases of Children* 6 (1913): 232–244.

23. Samuel McClintock Hamill, Howard C. Carpenter, and Thomas A. Cope, "A Comparison of the Pirquet, Calmette, and Moro Tuberculin Tests and Their Diagnostic Value," *Archives of Internal Medicine* 2 (1908): 405.

24. Albert Leffingwell, "Illustration of Human Vivisection," 1907.

## 2 EUGENICS AND THE DEVALUING OF INSTITUTIONALIZED CHILDREN

1. Edwin Black, *War against the Weak* (New York: Four Walls Eight Windows, 2003), p. 41.

2. The best article on the Letchworth Village castration is Paul Lombardo's "Tracking Chromosomes, Castrating Dwarves: Uninformed Consent and Eugenic Research," *Ethics & Medicine* 25, No. 3 (Fall 2009).

3. The condition known today as Down syndrome or Trisomy 21 was commonly referred to derisively as "mongolism" through much of the nineteenth and twentieth centuries; the term was used to describe those "idiotic and imbecilic" children of European or Caucasian background with "atavistic features" of a Mongolian from Asia. First described by John Langdon Haydon Down, a British physician, in 1866, the condition and whether it was hereditary or congenital would confound and stimulate scientific debate for generations.

4. Letter from Charles B. Davenport to Dr. C. S. Little, July 12, 1929. Papers of Charles B. Davenport at the American Philosophical Society (cited hereafter as C. B. Davenport Papers).

5. Ibid.

6. The child's father was deceased and his mother was said to be "of low mentality." Paul Lombardo describes her as "so lacking in comprehension that her consent would probably not be legally valid." "Tracking Chromosomes, Castrating Dwarves," p. 156.

7. Black, *War against the Weak,* p. 16.

8. William T. Belfield, "The Sterilization of Criminals and Other Defectives by Vasectomy," *Journal of the New Mexico Medical Society* (1909): 21–25.

9. Lewellys F. Barker, "The Importance of the Eugenic Movement and Its Relation to Social Hygiene," *Journal of the American Medical Association* 54 (1910): 2017–2022; and G. Frank Lydston, "Sex Mutilations in Social Therapeutics," *New York Medical Journal* 95 (1912): 677–685.

10. Phillip Reilly, "The Surgical Solution: The Writings of Activist Physicians in the Early Days of Eugenical Sterilization," *Perspectives in Biology and Medicine* 26, No. 4 (Summer 1983): 650.

11. Martin W. Barr, "Some Notes on Asexualization; with a Report of Eighteen Cases," *Journal of Nervous & Mental Disease* 51, No. 3 (March 1920): 232.

12. Leon F. Whitney, *The Case for Sterilisation* (London: John Lane, 1935), p. 76.

13. Ibid., p. 2.

14. Alexander Johnson, "To Eliminate the Defectives," December 28, 1932, pp. 1–3.

15. Whitney, *Case for Sterilisation,* p. 99. Whitney's book is a clear and powerful statement about the era and the attitudes of eugenicists. It is littered with demeaning references to those with disabilities and those placed in institutions involuntarily. For example, in addressing the elimination of such individuals, Whitney states, "Undoubtedly society would be better off without such, though the assertion has been made that we need them for our drudgery—for the dirty work of the world" (p. 100).

16. Black, *War against the Weak,* p. 99.

17. Ibid., p. 126.

18. From the turn of the century to the Great Depression, public institutions in America continued to grow in both numbers and population. By 1923, there were over 42,000 people in institutions, a number that would grow to 50,000 just three years later. By the mid-1930s, the population was over 81,000. James W. Trent Jr., *Inventing the Feeble Mind: A History of Mental Retardation in the United States* (Berkeley: University of California Press, 1994), p. 199.

19. Daniel J. Kevles, *In the Name of Eugenics: Genetics and the Uses of Human Heredity* (New York: Knopf, 1985), pp. 47, 53.

20. Ibid., p. 58.

21. Ibid., p. 55.

22. Black, *War against the Weak,* p. 94.

23. Lombardo, "Tracking Chromosomes, Castrating Dwarves."

24. Letchworth Village Cornerstone Program, June 14, 1933, pp. 149–164.

25. Letter from Charles B. Davenport to Henry H. Goddard, March 18, 1909. C. B. Davenport Papers.

26. Eugenic Record Office Questionnaire. C. B. Davenport Papers.

27. Letter from Henry Goddard to Charles B. Davenport, April 13, 1910. C. B. Davenport Papers.

28. Letter from Charles B. Davenport to H. H. Goddard, April 18, 1910. C. B. Davenport Papers, p. 3. Davenport would further mollify his concerned colleague about the potential for "exploit[ation]" by stating "the danger seems slight since people do not object so much to facts being known as to their relation to those facts being known. The families" being addressed in their publications "might just as well be on the planet Mars so far as any danger of their being identified by other people goes" (p. 4).

29. In adapting Binet's intelligence tests to English and the American educational landscape, Goddard coined a new term for the highest class of feebleminded individuals. The three

categories of IQ differentiation—idiot, imbecile, and moron—would remain part of institutional profiling for generations.

30. Letter from Henry Goddard to Charles B. Davenport, October 25, 1915. C. B. Davenport Papers.
31. Dozens of ERO questionnaires concerning university course work can be found in the C. B. Davenport Papers.
32. University of California Questionnaire. C. B. Davenport Papers.
33. Letter from Adolf Meyer to Charles B. Davenport, April 28, 1921. C. B. Davenport Papers.
34. Letter from F. Heiermann, S.J., to Harry B. Laughlin, February 16, 1920. C. B. Davenport Papers.
35. Walter Lippmann, "The Abuse of the Tests," *New Republic* (November 15, 1922): 297.
36. Quoted in Kevles, *In the Name of Eugenics*, p. 138.
37. Letter from E. G. Conklin to Charles B. Davenport, February 9, 1909. C. B. Davenport Papers.
38. Kevles, *In the Name of Eugenics*, p. 75.
39. Black, *War against the Weak*, p. 75.
40. Whitney, *Case for Sterilisation*, p. 195.
41. Kevles, *In the Name of Eugenics*, p. 110.
42. Ibid., p. 111.
43. Letter from Charles B. Davenport to C. S. Little, August 8, 1924. C. B. Davenport Papers.
44. Letter from Charles B. Davenport to C. S. Little, May 27, 1927. C. B. Davenport Papers.
45. Letter from C. S. Little to Charles B. Davenport, January 16, 1935. C. B. Davenport Papers.
46. Report of the Meeting of the Letchworth Village Research Council, January 15, 1936. C. B. Davenport Papers.
47. Letter from Charles B. Davenport to Elizabeth W. Buck, July 26, 1939. C. B. Davenport Papers.
48. Lombardo, "Tracking Chromosomes, Castrating Dwarves," p. 156.
49. Dr. Henry Heiman, "A Clinical and Bacteriological Study of the Gonoccus Neisser in the Male Urethra and in the Vulvo-vaginal Tract of Children," *Journal of Cutaneous and Genitourinary Diseases* 13 (1895): 385.

## 3  WORLD WAR II, PATRIOTISM, AND THE NUREMBERG CODE

1. David J. Rothman, *Strangers at the Bedside: A History of How Law and Bioethics Transformed Medical Decision Making* (New York: Basic Books, 1991), p. 36.
2. Ibid.
3. Jay Katz, "The Consent Principle of the Nuremberg Code," in *The Nazi Doctors and the Nuremberg Code*, eds. George J. Annas and Michael Grodin (New York: Oxford University Press, 1992), p. 228.
4. Jay Katz, *Final Report of the Advisory Committee on Human Radiation Experiments* (New York: Oxford University Press, 1996), p. 234.
5. Rothman, *Strangers at the Bedside*, p. 30.
6. Letter from Robert Ward to Dr. Joseph Stokes Jr., April 12, 1943. Papers of Dr. Joseph Stokes Jr., American Philosophical Society. Cited hereafter as J. Stokes Papers.
7. Ibid., p. 2.
8. Letter from Dr. Joseph Stokes Jr. to Dr. Paul de Kruif, November 15, 1935. J. Stokes Papers.
9. Letter from Dr. Joseph Stokes Jr. to Dr. Paul de Kruif, April 25, 1936. J. Stokes Papers.
10. Rothman, *Strangers at the Bedside*, p. 33.
11. Letter from Robert Ward to Dr. Stokes, April 7, 1943. J. Stokes Papers.
12. Annual report of the Commission on Neurotropic Virus Diseases, March 27, 1945, p. 4.
13. Ibid., p. 5.
14. Letter from S. Bayne-Jones to Captain John R. Neefe, August 4, 1944. J. Stokes Papers.
15. Elizabeth P. Maris, Geoffrey Rake, Joseph Stokes Jr., Morris P. Shaffer, and Gerald C. O'Neil, "Studies on Measles: The Results of Chance and Planned Exposure to Unmodified

Measles Virus in Children Previously Inoculated with Egg-Passage Measles Virus," *Journal of Pediatrics* 23, No. 6 (1943): 29.

16. Ibid., p. 17.
17. Letter from John R. Neefe to Maj. Walter P. Havens Jr. September 1, 1944. J. Stokes Papers.
18. Joseph Stokes Jr., "Mission to MTO USA and ETO USA at the request of the Surgeon General, U. S. Army, to Study the Prevention and Treatment of Epidemic Hepatitis." J. Stokes Papers.
19. Letter from Joseph Stokes Jr. to Brig. Gen. S. Bayne-Jones, September 21, 1944. J. Stokes Papers.
20. Letter from John R. Neefe to Brig. Gen. Stanhope Bayne-Jones, May 17, 1945. J. Stokes Papers.
21. Letter from Brig. Gen. S. Bayne-Jones to Capt. John R. Neefe, April 11, 1945. J. Stokes Papers.
22. Rothman, *Strangers at the Bedside,* p. 48.
23. Ibid.

## 4  IMPACT OF THE COLD WAR ON HUMAN EXPERIMENTATION

1. Author interviews with Karen Alves, October 26, 2001, and June 5, 2012.
2. Cerebral palsy is a neurologic disorder caused by damage to the motor control centers of the developing brain during pregnancy or childbirth resulting in limitations to movement and posture.
3. Esther M. Pond and Stuart A. Brody, *Evolution of Treatment Methods at a Hospital for the Mentally Retarded,* Department of Mental Hygiene, 1965, pp. 1–14. A copy is maintained by Bancroft Library, University of California, Berkeley.
4. Rebecca Leung, "A Dark Chapter in Medical History," *60 Minutes,* CBS, February 11, 2009.
5. Karen Alves would subsequently learn that LPNI #8732 stood for Langley Porter Neurological Center Patient Number 8732.
6. Neuropathology report on Mark Dal Molin, LPNI #8732, July 6, 1961. Supplied by his sister, Karen Alves.
7. Autopsy report for Mark Dal Molin, #19139, May 30, 1961.
8. Application for Research Grant, "An Etiological and Diagnostic Study of Cerebral Palsy," National Institutes of Health, October 20, 1952, p. 2.
9. Nathan Malamud, Hideo H. Itabashi, Jane Castor, and Harley B. Messinger, "An Etiologic and Diagnostic Study of Cerebral Palsy," *Journal of Pediatrics* 65, No. 2 (August 1964): 271.
10. Alves interviews.
11. "Hepatitis Studies—Pennhurst: Results, Tissue Culture Experiment #2," August 15, 1947, p. 1. Joseph S. Stokes Jr. papers at the American Philosophical Society. Cited hereafter as J. Stokes Papers.
12. Ibid., p. 2.
13. Ibid., p. 3.
14. Presidential Commission for the Study of Bioethical Issues, *"Ethically Impossible" STD Research in Guatemala from 1946 to 1948,* September 2011.
15. Distribution of Volunteers for Next Hepatitis Experiment, undated. J. Stokes Papers. Plans for Jaundice Experiments. American Philosophical Society.
16. David J. Rothman, *Strangers at the Bedside: A History of How Law and Bioethics Transformed Medical Decision Making* (New York: Basic Books, 1991), p. 51.
17. Ibid., p. 52.
18. Ibid., p. 53.
19. Ibid., p. 55.
20. "The Nuremberg Trial against German Physicians," *Journal of the American Medical Association* 135, No. 13 (November 29, 1947): 867.
21. Oral History Project of the Advisory Committee on Human Radiation Experiments, January 13, 1995, p. 2.

22. Authors' interview with Dr. Chester Southam, July 28, 1996.

23. Authors' interview with Dr. A. Bernard Ackerman, September 9, 1996.

24. Letter from Joseph Stokes Jr. to Col. William B. Stone, October 15, 1947. J. Stokes Papers.

25. Letter from Joseph Stokes Jr. to Dr. Colin MacLeod, February 11, 1948. J. Stokes Papers.

26. Letter from C. J. Watson to Lt. Col. Frank L. Bauer, April 12, 1948. Papers of Joseph Stokes, Jr., American Philosophical Society.

27. Letter from Dr. John R. Paul to Dr. Joseph Stokes Jr., February 18, 1948. J. Stokes Papers.

28. Letter from Dr. Joseph Stokes Jr. to Dr. Colin MacLeod, February 11, 1948. J. Stokes Papers.

29. Dr. Richard B. Capps, "Proposed Studies on Liver Disease," March 10, 1951. J. Stokes Papers.

30. The membership list of the Sub-Committee on the Allocation of Volunteers of the Armed Forces Epidemiological Board included Dr. Irving Gordon of the Division of Laboratories & Research for the Department of Health; Dr. Roderick Murray of the National Institutes of Health; Dr. Cecil Watson, director of the Commission on Liver Disease and professor at the University of Minnesota; Dr. Colin MacLeod, president of the AFEB; and Colonel A. J. Rapalski, a key administrator of the AFEB at the Office of the Surgeon General. J. Stokes Papers.

31. Letter from Frank L. Bauer to Dr. Joseph Stokes Jr., April 26, 1950. J. Stokes Papers.

32. Letter from Dr. Joseph Stokes Jr. to Dr. Cecil J. Watson, February 8, 1951. J. Stokes Papers.

33. Armed Forces Epidemiological Board, Office of Medical History, US Army Medical Department.

34. Letter from Dr. Joseph Stokes Jr. to Dr. Cecil J. Watson, February 4, 1953. J. Stokes Papers.

35. Letter from Dr. Joseph Stokes Jr. to Dr. Cecil J. Watson, February 5, 1953. J. Stokes Papers.

36. Letter from Alex M. Burgess to Dr. Joseph Stokes Jr. December 10, 1945. J. Stokes Papers.

37. Memorandum from Andy Burgess to Homer L. Morris on Volunteer Experimental Subjects. J. Stokes Papers, p. 2.

38. Susan E. Lederer, *Subjected to Science: Human Experimentation in America before the Second World War* (Baltimore: Johns Hopkins University Press, 1997), p. 137.

39. Letter from Dr. Colin M. MacLeod to Dr. Herman E. Hillsboe, December 3, 1952. J. Stokes Papers, p. 1.

40. Ibid., p. 2.

41. Joseph Stokes Jr., "A Clarification of the Question of Ethical Responsibility in the Exposure of Human Beings to Certain Infectious Agents," February 4, 1953, p. 1. J. Stokes Papers.

42. Ibid., p. 4.

43. Ibid., p. 6.

44. Letter from Dr. Roderick Murray to Dr. Joseph Stokes Jr., November 13, 1952. J. Stokes Papers.

45. Pope Pius XII, "The Moral Limits of Medical Research and Treatment," September 26, 1952, in the Papers of Joseph Stokes Jr. at the American Philosophical Society, pp. 1–2.

46. Unsigned and undated memo on the subject of prisoner compulsion. Papers of Joseph Stokes, Jr., American Philosophical Society.

47. Both *Acres of Skin: Human Experiments at Holmesburg Prison; A True Story of Abuse and Exploitation in the Name of Medical Science* (New York: Routledge, 1998) and *Sentenced to Science: One Black Man's Story of Imprisonment in America* (University Park: Pennsylvania State University Press, 2007) by Allen M. Hornblum provide a historical accounting of prison research in America and the reasons inmates chose to participate in experimental studies.

48. Department of Defense Resolution concerning House of Delegates of the American Medical Association. J. Stokes Papers.

49. Letter from Dr. Joseph Stokes Jr. to Dr. Paul Havens, November 10, 1952. J. Stokes Papers, p. 1.

50. Letter from Nate Hazeltine to Dr. Colin Macleod, February 10, 1953. J. Stokes Papers.

51. Letter from Dr. Joseph Stokes Jr. to Harry Von Bulow, Jr., July 7, 1954. J. Stokes Papers.

52. Letter from H. Von Bulow to Dr. Joseph Stokes Jr., July 12, 1954. J. Stokes Papers.

53. Letter from H. Von Bulow to Dr. Joseph Stokes Jr., August 31, 1954. J. Stokes Papers.

54. Letter from Dr. Joseph Stokes Jr. to Dr. Herald Cox, August 24, 1954. J. Stokes Papers.
55. Letter from Herald R. Cox to Dr. Joseph Stokes Jr., August 26, 1954. J. Stokes Papers.
56. Letter from Dr. Joseph Stokes Jr. to Mrs. Rose Antman, September 28, 1955. J. Stokes Papers.
57. Paul Starr, *The Social Transformation of American Medicine: The Rise of a Sovereign Profession and the Making of a Vast Industry* (New York: Basic Books, 1982), p. 335.
58. Ibid., p. 336.
59. J. Edgar Hoover, *Masters of Deceit: The Story of Communism in America and How to Fight It* (New York: Henry Holt, 1958), p. 331.
60. Paul A. Offit, *The Cutter Incident: How America's First Polio Vaccine Led to the Growing Vaccine Crisis* (New Haven, CT: Yale University Press, 2005), p. 84.
61. Christopher Simpson, *Blowback: The First Full Account of America's Recruitment of Nazis and Its Disastrous Effect on the Cold War, Our Domestic and Foreign Policy* (New York: Collier Books, 1988), p. 3.
62. Allen Weinstein and Alexander Vassiliev, *The Haunted Wood: Soviet Espionage in America— The Stalin Era* (New York: Modern Library, 1999), p. 300.
63. Richard Rhodes, *Dark Sun: The Making of the Hydrogen Bomb* (New York: Simon & Schuster, 1995), p. 260.
64. James S. Ketchum, *Chemical Warfare Secrets Almost Forgotten: A Personal Story of Medical Testing of Army Volunteers* (Santa Rosa, CA: ChemBook, 2006), p. 3.
65. Advisory Committee on Human Radiation Experiments, *Final Report of the Advisory Committee on Human Radiation Experiments* (New York: Oxford University Press, 1996), p. 2.
66. Ibid.
67. John D. Marks, *The Search for the Manchurian Candidate: The CIA and Mind Control; The Secret History of the Behavioral Sciences* (New York: W. W. Norton, 1979), p. vii.
68. Authors' interview with Dr. James Ketchum, July 23, 2012.
69. Ibid., p. 247.
70. Authors' interview with Dr. Enoch Callaway, October 30, 2010.
71. Marks, *The Search for the Manchurian Candidate*, p. 25.
72. Ibid., p. 63.
73. Bufotenine is a psychedelic neurotransmitter found in certain toads, mushrooms, and plants.
74. Marks, *The Search for the Manchurian Candidate*, p. 142.
75. Ibid., p. 150.
76. Harvey Weinstein, *Father, Son and CIA* (Halifax: Goodread Biographies, 1990), p. 294.
77. Marks, *The Search for the Manchurian Candidate*, p. 18.
78. Alton Chase, *Harvard and the Unabomber: The Education of an American Terrorist* (New York: Norton, 2002).
79. Henry K. Beecher, "Ethics and Clinical Research," *New England Journal of Medicine* 274, No. 24 (June 16, 1966): 1354–1360.
80. David J. Rothman, "Ethics and Human Experimentation: Henry Beecher Revisited," *New England Journal of Medicine* 317, No. 19 (November 5, 1987): 1195.
81. Beecher, "Ethics and Clinical Research," p. 1354.
82. Rothman, "Ethics and Human Experimentation," p. 1196.
83. Maurice Pappworth, *Human Guinea Pigs* (Boston: Beacon Press, 1967).

## 5 VACCINES

1. Interview with Pat Clapp, August 5–7, 2012.
2. Henry W. Pierce, "State Halts Use of Retarded as Guinea Pigs," *Pittsburgh Post-Gazette*, April 11, 1973; Henry W. Pierce, "State Kills Testing of Meningitis Shots," *Pittsburgh Post-Gazette*, April 12, 1973; Delores Frederick, "Experiments on Humans Denied Here," *Pittsburgh Press*, April 12, 1973.
3. Pierce, "State Kills Testing of Meningitis Shots."
4. Pierce, "State Halts Use of Retarded as Guinea Pigs."
5. Pierce, "State Kills Testing of Meningitis Shots."
6. Ibid.

7. *Prince v. Massachusetts,* 321 US 158 (1944).

8. Quoted in Delores Frederick, "Courts, Not Parents, Have Word on Retarded Tests, Wecht Says," *Pittsburgh Press,* April 12, 1973.

9. "Retarded at Hamburg State Drugged," *Pittsburgh Post-Gazette,* April 20, 1973, p. 13.

10. Gabriel Ireton, "Polk Chaplin Defends Use of Patient Pens," *Pittsburgh Post-Gazette,* June 21, 1973.

11. "Designed Cage, Doctor Admits."

12. Letter from Helene Wohlgemuth to James H. McClelland, April 16, 1973. From Pat Clapp's personal files.

13. Paul Offit, *The Cutter Incident* (New Haven, CT: Yale University Press, 2005), p. 35.

14. David Oshinski, *Polio: An American Story* (New York: Oxford University Press, 2005), p. 151.

15. Jonas E. Salk, Harold E. Pearson, Philip N. Brown, and Thomas Francis Jr., "Protective Effect of Vaccination against Induced Influenza B," *Journal of Clinical Investigation* 24, No. 4 (July 1945): 547–553.

16. Quoted in Oshinski, *Polio.*

17. Superintendents of orphanages, mental institutions, and prisons had enormous power during the first seventy-five years of the twentieth century. They ran the facilities like personal fiefdoms and often allowed medical professionals to use patients or prisoners for various scientific projects. In many cases, governing authorities had little knowledge of what was going on inside the walls of an institution. It was not unusual for superintendents to shake hands on a deal that would allow physicians and others exclusive access and use of a facility that would last many years, sometimes decades. See Allen M. Hornblum, *Acres of Skin: Human Experiments at Holmesburg Prison; A True Story of Abuse and Exploitation in the Name of Medical Science* (New York: Routledge, 1998).

18. Letter from Dr. Gale H. Walker to Honorable William C. Brown, February 4, 1952, Salk Archives, University of California, San Diego (hereafter UCSD), p. 1.

19. Letter from Hilding Bengs to Dr. Gale H. Walker, March 18, 1952, Salk Archives, UCSD.

20. Irena Koprowska, *A Woman Wanders through Life and Science* (Albany: State University of New York Press, 1997), p. 298.

21. Interview with Dr. Hilary Koprowski, September 14, 2009.

22. Koprowska, *A Woman Wanders through Life and Science,* p. 298.

23. Quoted in ibid., p. 299.

24. Koprowski interview.

25. Quoted in Saul Benison, *Tom Rivers: Reflections on a Life in Medicine and Science* (Cambridge, MA: MIT Press, 1967), p. 465.

26. Ibid.

27. Ibid., p. 467.

28. "Poliomyelitis: A New Approach," *Lancet,* March 15, 1952, p. 552.

29. Howard A. Howe, "Criteria for the Inactivation of Poliomelitis," Committee on Immunization, December 4, 1951, meeting. Mandeville Special Collections Library, UCSD, p. 3.

30. Howard A. Howe, "Antibody Response of Chimpanzees and Human Beings to Formalin-Inactivated Trivalent Poliomyelitis Vaccine," *American Journal of Epidemiology* 56, No. 3 (1952): 265–286.

31. "The Prophylaxis of Poliomyelitis" is a transcript of the April 30, 1936, meeting in Baltimore, Maryland, p. 4.

32. N. Paul Hudson, Edwin H. Lennette, and Francis B. Gordon, "Factors of Resistance in Experimental Poliomyelitis," *Journal of the American Medical Association* 106, No. 24 (June 1936): 1.

33. "The Prophylaxis of Poliomyelitis," p. 15.

34. Ibid., p. 17.

35. Ibid.

36. Ibid., p. 33.

37. Morris Schaeffer, "William H. Park: His Laboratory and His Legacy," *American Journal of Public Health* 75, No. 11 (November 1985): 1301.

38. "New Infantile Paralysis Vaccine Is Declared to Immunize Children," *New York Times,* August 18, 1934; "Will Give Children Paralysis Vaccine," *New York Times,* August 21, 1934.
39. Schaeffer, "William H. Park."
40. "Questions Safety of Paralysis Virus," *New York Times,* October 9, 1935.
41. Ibid.
42. "Dr. Brodie Upholds Paralysis Vaccine," *New York Times,* November 3, 1935.
43. Schaeffer, "William H. Park."
44. Henry K. Beecher, "Ethics and Clinical Research," *New England Journal of Medicine* 274, No. 24 (June 16, 1966): 1354–1360.
45. "Was Dr. Krugman Justified," *World Medical News,* October 15, 1971, p. 29.
46. Stephen Goldby, "Experiments at the Willowbrook State School," *Lancet,* April 10, 1971, p. 749.
47. Ibid.
48. "Was Dr. Krugman Justified," p. 29.
49. Saul Krugman, "The Willowbrook Hepatitis Studies Revisited: Ethical Aspects," *Reviews of Infectious Diseases* 8, No. 1 (January-February 1986): 157.
50. Robert Ward, Saul Krugman, Joan P. Giles, and Milton A. Jacobs, "Endemic Viral Hepatitis in an Institution Epidemiology and Control," undated document ca. 1956, New York University Medical School Archives, p. 3. Hereafter NYU Archives.
51. The stool virus was developed in the following manner. "A 20% aqueous suspension was prepared from stools of 6 patients in the first 8 days of observed jaundice. Three centrifugations were carried out, 1 at 2000 RPM and 2 at 8500 RPM for 1 hour. The supernate was heated at 56 degrees C for 1/2 hour and treated with penicillin 1000 units and chloramphenicol 100 ug/ml. before sterility was achieved. This sterile suspension was inoculated into 5 monkeys (1 cc intracerebrally), 47 suckling mice, Hela and monkey kidney tissue cultures. No changes occurred in any of these test objects and the monkeys' spinal cords and brain stems showed no evidence of poliomyelitis lesions." Ibid., p. 5.
52. Ibid., p. 6.
53. Ibid., p. 7.
54. Letter from Saul Krugman to Harold H. Berman, January 27, 1960. Papers of Saul Krugman, NYU Archives.
55. Letter from Robert Ward to Dr. Harold H. Berman, March 29, 1957. Papers of Saul Krugman, NYU Archives.
56. Ibid., pp. 2–4.
57. Letter from Saul Krugman to Harold H. Berman, April 27, 1959. Papers of Saul Krugman, NYU Archives.
58. Letter from Alfred M. Prince to Dr. Saul Krugman, November 29, 1960. Papers of Saul Krugman, NYU Archives.
59. Letter from Saul Krugman to Capt. Alfred M. Prince, December 14, 1960. Papers of Saul Krugman, NYU Archives.
60. Letter from Saul Krugman to Jack Hammond, August 2, 1965. Papers of Saul Krugman, NYU Archives.
61. Letter from Saul Krugman to Jack Hammond, July 12, 1966. Papers of Saul Krugman, NYU Archives.
62. Letter from Jack Hammond to Saul Krugman, November, 24, 1967. Papers of Saul Krugman, NYU Archives.
63. Lawrence K. Altman, "Immunization Is Reported in Serum Hepatitis Tests," *New York Times,* March 24, 1971.
64. Letter from Stanley A. Plotkin to Mr. Robert Morrow, March 15, 1972. Papers of Saul Krugman, NYU Archives.
65. Stanley A. Plotkin, David Cornfield, and Theodore H. Ingalls, "Studies of Immunization with Living Rubella Virus Trials in Children with a Strain Cultured from an Aborted Fetus," *American Journal of Disability and Children* 110 (October 1965): 382.
66. Ibid.

67. Stanley A. Plotkin, J. D. FarQuhar, M. Katz, and C. Hertz, "Further Studies of an Attenuated Rubella Strain Grown in WI-38 Cells," *American Journal of Epidemiology* 89, No. 2 (September 1968): 238.

68. Letter from Richard Capps to Dr. Joseph Stokes Jr., November 30, 1950. Papers of Dr. Joseph Stokes Jr. at the American Philosophical Society.

69. Ibid. Handwritten comments at top of page.

70. It is difficult to determine if the St. Vincent experiment was carried out; no journal articles or additional correspondence describe its results. The letters that do exist disclose that a good deal of thought went into the issue and the intended plan.

71. William H. Wilder, "Report of the Committee for the Study of the Relation of Tuberculosis to Diseases of the Eye," *Journal of the American Medical Association* 55, No. 1 (June 1910): 21.

72. Samuel McC. Hamill, Howard Childs Carpenter, and Thomas Cope, "A Comparison of the Von Pirquet, Calmette and Moro Tuberculin Tests and Their Diagnostic Value," *Archives of Internal Medicine* 2, No. 5 (December 1908): 419.

73. Susan E. Lederer, *Subjected to Science* (Baltimore: Johns Hopkins University Press, 1995), p. 80.

74. Louis M. Warfield, "The Cutaneous Tuberculin Reaction," *Journal of the American Medical Association* 50, No. 9 (February 1908): 688.

75. L. Emmett Holt, "A Report Upon One Thousand Tuberculin Tests in Young Children," *Archives of Pediatrics* (January 1909): 3.

76. Ibid., pp. 2, 7.

77. Louis W. Sauer and Winston H. Tucker, "Immune Responses to Diptheria, Tetanus, and Pertussis, Aluminum Phosphate Absorbed" *Journal of Public Health* 44, No. 6 (1954): 785.

78. Ibid.

79. Johannes Ipsen, Jr., "Bio-Assay of Four Tetanus Toxoids (Aluminum Precipitated) in Mice, Guinea Pigs, and Humans," *Journal of Immunology* 70, No. 4 (1953): 426–434.

80. Ibid., p. 433.

81. Stephen Millan, John Maisel, C. Henry Kempe, Stanley Plotkin, Joseph Pagano, and Joel Warren, "Antibody Response of Man to Canine Distemper Virus," *Journal of Bacteriology* 79, No. 4 (1960): 618.

82. Werner Henle, Gertrude Henle, and Joseph Stokes Jr., "Demonstration of the Efficacy of Vaccination Against Influenza Type A by Experimental Infection of Human Beings," *Journal of Immunology* 46 (1943): 163.

83. Ibid., p. 166.

84. Joseph Stokes Jr., Robert E. Weibel, Eugene B. Buynak, and Maurice R. Hilleman, "Live Attenuated Mumps Virus Vaccine," *Pediatrics* 39, No. 3 (1967): 363.

85. George T. Harrell, S. F. Horne, Jerry K. Aikawa, and Nancy J. Helsabeck, "Trichinella Skin Tests in an Orphanage and Prison: Comparison with Serologic Tests for Trichinosis and with the Tuberculin Reaction," *Journal of Clinical Investigation* 26, No. 1 (January 1947): 64.

## 6  SKIN, DIETARY, AND DENTAL STUDIES

1. Interview with Margaret Grey Wood, February 2, 1996.

2. Interview with Paul Gross, January 22, 1996.

3. Kligman's many ringworm articles caused a member of the Holmesburg Prison medical staff to ask if he would take a look at the athlete's foot problem at the institution. Kligman visited the jail and immediately viewed it as "a fertile field" of investigatory opportunity. He performed experiments on the inmates for the next twenty-four years.

4. Albert M. Kligman, "The Pathogenesis of *Tinea capitis* due to *Microsporum audouini* and *Microsporum canis*," *Journal of Investigative Dermatology* 18, No. 3 (March 1952): 231.

5. Ibid., p. 246.

6. Albert M. Kligman and W. Ward Anderson, "Evaluation of Current Methods for the Local Treatment of *Tinea capitis*," *Journal of Investigative Dermatology* 16 (March 1951): 162.

7. Quoted in Allen M. Hornblum, *Acres of Skin: Human Experiments at Holmesburg Prison; A True Story of Abuse and Exploitation in the Name of Medical Science* (New York: Routledge, 1998), p. 37.

8. Botho F. Felden, "Epilation with Thallium Acetate in the Treatment of Ringworm of the Scalp of Children," *Archives of Dermatology and Syphilology* 17, No. 2 (February 1928): 185.

9. Robert E. Swain and W. G. Bateman, "The Toxicity of Thallium Salts," *Journal of Biological Chemistry* (December 9, 1909): 147.

10. G. B. Dowling, "The Treatment of Ringworm of the Scalp by Thallium Depilation," *British Medical Journal* 2, No. 3475 (August 13, 1927): 261.

11. Felden, p. 192.

12. Howard Ticktin, "Hepatic Dysfunction and Jaundice in Patients," *New England Journal of Medicine* 267, No. 19 (1962): 964–968.

13. Isaac A. Abt, "Spontaneous Hemorrhages in Newborn Children," *Journal of the American Medical Association* 40, No. 5 (January 1903): 290.

14. Ibid.

15. Ibid., p. 291.

16. Joann G. Elmore and Alvan R. Feinstein, "Joseph Goldberger: An Unsung Hero of American Clinical Epidemiology," *Annals of Internal Medicine* 121, No. 5 (September 1994): 372.

17. Hornblum, *Acres of Skin*, p. 77.

18. Quoted in Alan M. Kraut, *Goldberger's War* (New York: Hill and Wang, 2003), p. 105.

19. Quoted in ibid., p. 106.

20. Donald J. Barnes, "Effect of Evaporated Milk on the Incidence of Rickets in Infants," *Journal of the Michigan State Medical Society* 31 (1932): 397.

21. Victor A. Najjar and L. Emmett Holt, "The Biosynthesis of Thiamine in Man and Its Implications in Human Nutrition," *Journal of the American Medical Association* 123, No. 11 (November 1943): 683.

22. Victor A. Najjar, George C. Johns, George C. Mediary, Gertrude Fleischmann, and L. Emmett Holt, "The Biosynthesis of Riboflavins," *Journal of the American Medical Association* 126, No. 6 (October 1944): 357–358.

23. Pauline B. Mack, Alice Knapper Milsom, and Paul L. Carney, "A Study of Two Levels of Bread Enrichment in Children's Diet," *Monographs of the Society for Research in Child Development* 18, No. 2 (1953): 1–92.

24. Herbert I. Goldman, Samuel Karelitz, Eli Selfter, Hedda Acs, and Norman B. Schell, "Acidosis in Premature Infants Due to Lack of Lactic Acid," *Pediatrics* 27 (1961): 928.

25. Ibid., p. 929.

26. Norman B. Schell, Samuel Karelitz, and Bernard S. Epstein, "Radiographic Study of Gastric Emptying in Premature Infants," *Journal of Pediatrics* 62, No. 3 (March 1963): 343.

27. Basil G. Bibby, "A Test of the Effect of Fluoride-containing Dentifrices," *Journal of Dental Research* 24, No. 6 (1945): 297–303.

28. Julian D. Boyd and Kenneth E. Wessels, "Epidemiologic Studies in Dental Caries: The Interpretation of Clinical Data Relating to Caries Advance," *American Journal of Public Health* 41 (August 1951): 978.

29. Ibid., p. 979.

30. Ibid., p. 983.

31. R. L. Glass and S. Fleisch, "Diet and Dental Caries: Dental Caries Incidence and the Consumption of Ready-to-Eat Cereals," *Journal of the American Dental Association* 88, No. 4 (April 1974): 807–813; N. H. Rowe, R. H. Anderson, and L. A. Wanninger, "Effects of Ready-to-Eat Breakfast Cereals on Dental Caries Experience in Adolescent Children," *Journal of Dental Research* 53, No. 1 (January 1974): 33–36; Thomas J. Hill, John Sims, and Millicent Newman, "The Effect of Penicillin Dentifrice on the Control of Dental Caries," *Journal of Dental Research* 32 (1953): 448–452.

32. A. D. Steinberg, S. O. Zimmerman, and M. L. Bramer, "The Lincoln Dental Caries Study: The Effect of Acidulated Carbonated Beverages on the Increase of Dental Caries," *Journal of the American Dental Association* 85, No. 1 (1972): 81–89.

## 7 RADIATION EXPERIMENTS ON CHILDREN

1. Authors' interviews with Gordon Shattuck. Numerous interviews were conducted with Shattuck in Waltham and Beverly, Massachusetts, as well as by phone, between July 2010 and August 2011.

2. Case Record Folder for Gordon C. Shattuck, Walter E. Fernald State School, April 23, 1947, p. 1.

3. Walter E. Fernald died on the evening of November 27, 1924. The Massachusetts legislature declared on July 1, 1925, that the Massachusetts School for the Feeble-minded would become the Walter E. Fernald State School. Guy Pratt Davis, *What Shall the Public Schools Do for the Feeble-Minded?* (Cambridge, MA: Harvard University Press, 1927).

4. Marie E. Daly, "History of the Walter E. Fernald Development Center," p. 2.

5. Quoted in Michael D'Antonio, *The State Boys Rebellion* (New York: Simon & Shuster, 2004), p. 19.

6. Shattuck interviews.

7. Authors' interviews with Charles Dyer. Numerous interviews were conducted with Dyer in Waltham and Beverly, Massachusetts, as well as by phone, between July 2010 and August 2011.

8. Authors' interviews with Austin LaRocque. Numerous interviews were conducted with LaRocque in Waltham and Beverly, Massachusetts, as well as by phone, between September 2009 and August 2011.

9. Shattuck interviews.

10. Clinical Data and Further Experimental Details—Experiment No. 1, March 28, 1950. Included in report by the Task Force on Human Subject Research, "A Report on the Use of Radioactive Materials in Human Subject Research that Involved Residents of State-Operated Facilities within the Commonwealth of Massachusetts from 1943 to 1973," April 1994. (Hereafter cited as Massachusetts Task Force Report.)

11. Dyer interviews.

12. In retrospect, a misstep by Dyer leading to his falling might have precipitated an investigation leading to an examination and termination of the MIT study. In point of fact, the medical experiments at Fernald and Wrentham would continue for another fifteen years.

13. LaRocque interviews.

14. Scott Allen, "Radiation Used on Retarded," *Boston Globe,* December 26, 1993.

15. Ibid.

16. Shattuck interviews.

17. Radiation abuse stories would saturate media outlets around the nation.

18. "Fernald's Atomic Cafe," *Boston Globe,* December 29, 1993.

19. Eileen Welsome, "Plutonium Experiment," *Albuquerque Tribune,* November 15–17, 1993.

20. Philip Campbell, Department of Mental Retardation memo, April, 15, 1994, pp. 29–37. From Massachusetts Task Force Report.

21. Massachusetts Task Force Report.

22. Quoted in ibid., p. 33.

23. Ibid., pp. 36–38.

24. Ibid., p. 35.

25. Quotations from Letter from Robert S. Harris to Dr. Malcolm Farrell, December 19, 1945. From Massachusetts Task Force Report, pp. 1, 2, 4–5.

26. Letter from Clifton T. Perkins to Malcolm J. Farrell, December 27, 1945. From Massachusetts Task Force Report.

27. "Report of Progress in Research," Report II, July 1, 1945, to June 30, 1946, Nutritional Biochemistry Laboratories, Department of Food Technology, Massachusetts Institute of Technology, pp. 11–12.

28. Letter from Robert S. Harris to Dr. Clemens E. Benda, April 13, 1949, "Outline of Proposed Experiment to Determine the Absorption of Calcium by Children and the Effect of Phytates upon Absorption," Massachusetts Task Force Report, p. 1.

29. Ibid., p. 2.

30. Letter from Clemens E. Benda to Paul Aberhold, May 18, 1949. Massachusetts Task Force Report, p. 1.

31. From the Papers of Clemens Ernst Benda, M.D., 1898–1975, Countway Medical Library, Harvard Medical School.

32. Regrettably, much of Benda's personal correspondence and research notes are at Countway Medical Library and have been sealed and are unavailable to the public and medical historians. The authors will continue to pursue access to these files.

33. Letter from S. Allan Lough to Dr. Robley D. Evans, September 28, 1949. Massachusetts Task Force Report, p. 1.

34. Letter from S. Allan Lough to Dr. Clemens E. Benda, November 3, 1949. Massachusetts Task Force Report, p. 1.

35. Ibid.

36. Ibid., p. 2.

37. Letter from Malcolm J. Farrell to Parents, November 2, 1949. Massachusetts Task Force Report, p. 1.

38. Letter from Robert S. Harris to Clemens E. Benda, May 1, 1953. Massachusetts Task Force Report, p. 1.

39. Ibid.

40. Ibid., p. 2.

41. Quotations from Letter from Clemens E. Benda to Dear Parent, May 28, 1953. Massachusetts Task Force Report, p. 1.

42. In a letter to a parent, Benda said he had learned of the "plan to take your son [name redacted] on vacation on July 1st." Benda informed the child's mother that her son was "cooperating in a science test" that was deemed "extremely important" and that he was needed at the institution for further tests. He then requested that the family's vacation be changed "from Wednesday, July 1st, to Friday, July 3rd." Letter from Clemens Benda to Mrs. [name redacted], June 29, 1953. Massachusetts Task Force Report.

43. For example, one child from Ward 22 was said to have been born in 1940, had a mental age of 82 an IQ of 55, and weighed 140 pounds. Science Club Boys, 6/5/53. Massachusetts Task Force Report.

44. In the early to mid-1950s, Fernald hosted a broad array of research projects. Benda admitted in a 1953 annual report that the school's research department was involved in a "larger field of investigations than in previous years" and now included work on "neuropathological and biochemical aspects of mongolism," work with Harvard's Department of Food Research" on metabolism studies, feeding and mating experiments on rats and mice, along with the MIT food metabolism research. Annual report of the Research Laboratory, Walter E. Fernald State School, July 1952 June 1953. Massachusetts Task Force Report.

45. Letter from Clemens E. Benda to Atomic Energy Commission, September 29, 1953. Massachusetts Task Force Report, p. 1.

46. Ibid.

47. Felix Bronner, Clemens E. Benda, Robert S. Harris, and Joseph Kreplick, "Calcium Metabolism in a Case of Gargoylism Studied with the Aid of Radiocalcium," *Journal of Clinical Investigation* 37 (1958): 139–147.

48. Letter from Jack R. Ewalt to Superintendent, June 1, 1956. Massachusetts Task Force Report, p. 1.

49. Quoted in Allen, "Radiation Used on Retarded."

50. Authors' interviews with Constantine Maletskos by telephone on February 22, 2011, and at Spaulding Rehabilitation Hospital, Boston, April 28, 2011.

51. The calcium studies were published by F. Bronner, R. S. Harris, C. J. Maletskos, and C. E. Benda, "Studies in Calcium Metabolism. Effect of Food Phytates on Calcium Uptake in Children on Low Calcium Breakfasts," *Journal of Nutrition* 54 (1954): 523–542; F. Bronner, R. S. Harris, C. J. Maletskos, and C. E. Benda, "Studies in Calcium Metabolism. Effects of Food Phytates on Calcium 45 Uptake in Children on Moderate Calcium Breakfasts," *Journal of Nutrition* 59 (1956): 393–406; and F. Bronner, R. S. Harris, C. J. Maletskos, and C. E. Benda, "Studies in Calcium Metabolism. The Fate of Intravenously Injected Radiocalcium in Human Beings," *Journal of Clinical Investigation* 35 (1956): 78–88.

52. Henry K. Beecher, "Ethics and Clinical Research," *New England Journal of Medicine* 274 (June 16, 1966): 24.

53. G. S. Kurland, J. Fishman, M. W. Hamolsky, and A. S. Freedberg, "Radioisotope Study of Thyroid Function in 21 Mongoloid Subjects, Including Observations in 7 Parents," *Journal of Clinical Endocrinology & Metabolism* 17, No. 4 (April 1957): 552.

54. Massachusetts Task Force Report, p. 14.

55. Ibid.

56. Ibid., p. 18. The study was published in 1962 by K. M. Saxena, E. M. Chapman, and C. V. Pryles, "Minimal Dosage of Iodide Required to Suppress Uptake of Iodine-131 by Normal Thyroid," *Science* 138 (1962): 430–431.

57. K. M. Saxena and C. V. Pryles, "Thyroid Function in Mongolism," *Journal of Pediatrics* 67 (1965): 363–370.

58. C. E. Benda, C. J. Maletskos, J. C. Hutchinson, and E. B. Thomas, "Studies of Thyroid Function in Myotonia Dystrophica," *American Journal of Medical Sciences* 228 (1954): 668–672.

59. Memo from Dr. Roy Shore to the Task Force, March 5, 1994, p. 1; and Memo from Joseph L. Lyon to Task Force. Massachusetts Task Force Report, p. 1.

60. Letter from Dr. Brian Macmahon to Doe West, February 24, 1994. Massachusetts Task Force Report, p. 1.

61. Authors' phone interview with Joseph Almeida, September 7, 2009.

62. Authors' phone interview with Sandra Marlow, April 10, 2010.

63. Authors' interviews with Doe West in Boston, April 28, 2010, and April 29, 2011.

64. George H. Lowery, William H. Beierwalters, Isadore Lampe, and Henry J. Gomberg, "Radioactive Uptake Curve in Human: Studies in Children," *Pediatrics* 4 (1949): 628.

65. Van Middlesworth, "Radioactive Iodide Uptake of Normal Newborn Infants," *American Journal of Diseases in Children* 38 (1954): 439.

66. Ibid., p. 441.

67. Edgar E. Martmer, Kenneth E. Corrigang, Harold P. Chareneau, and Allen Sosin, "A Study of the Uptake of Iodine by the Thyroid of Premature Infants," *Pediatrics* 17 (1955): 507.

68. Richard E. Ogborn, Ronald E. Waggener, and Eugene VanHove, "Radioactive-Iodine Concentration in Thyroid Glands of Newborn Infants," *Pediatrics* 28 (1960): 771.

69. Krishna M. Saxena, Earle M. Chapman, and Charles V. Pryles, "Minimal Dosage of Iodide Required to Suppress Uptake of Iodide-131 by Normal Thyroid," *Science* 138 (1962): 430.

## 8  PSYCHOLOGICAL TREATMENT

1. Authors' interviews with Ted Chabasinski, March 15, 2010, and October 26, 2011.

2. Unpublished manuscript by Ted Chabasinski provided to authors.

3. "The Papers of Lauretta Bender," Brooklyn College Library Archive.

4. Ibid.

5. In 1938, Bender wrote "A Visual Motor Gestalt Test and Its Clinical Use," (American Orthopsychiatric Association Monograph, 1938), which reproduced nine figures derived from the work of Gestalt psychologist Max Werthheimer. The Bender-Gestalt Test would become one of the most used tests by clinical psychologists to measure perceptual motor skills and perceptual motor development along with neurological intactness.

6. Lauretta Bender, "Theory and Treatment of Childhood Schizophrenia," *Acta Paedopsychiatrica* 34 (1968): 301.

7. Joel Braslow, *Mental Ills and Bodily Cures* (Berkeley: University of California Press, 1997), p. 96.

8. Ibid., p. 100.

9. Ibid., p. 101.

10. Bender, "Theory and Treatment of Childhood Schizophrenia."

11. Ibid.

12. Lauretta Bender, "One Hundred Cases of Childhood Schizophrenia Treated with Electric Shock," *Transactions of the American Neurologic Association* 762 (1947): 168.

13. Braslow, *Mental Ills and Bodily Cures,* p. 98.
14. Quoted in E. Shorter and D. Healy, *Shock Therapy* (New Brunswick, NJ: Rutgers University Press, 2007), p. 86.
15. Quoted in ibid., p. 87.
16. Ibid., p. 137.
17. Jack El-Hai, *The Lobotomist: A Maverick Medical Genius and His Tragic Quest to Ride the World of Mental Illness* (Hoboken, NJ: John Wiley, 2007), p. 107.
18. Ibid., p. 1.
19. Walter Freeman, "The Religion of Science." Papers of Walter Freeman, George Washington University Archives (cited hereafter as Freeman Papers).
20. Quoted in El-Hai, *The Lobotomist,* p. 180.
21. Quoted in ibid., p. 248.
22. Quoted in ibid., p. 91.
23. Edward Shorter, *A History of Psychiatry: From the Era of the Asylum to the Age of Prozac* (New York: John Wiley, 1997), p. 227.
24. "Lobotomy Disappointment," *Newsweek,* December 12, 1949, p. 51.
25. Quoted in El-Hai, *The Lobotomist,* p. 199.
26. Ibid., pp. 174–175.
27. Walter Freeman and James W. Watt, "Schizophrenia in Childhood," February 26, 1947. Freeman Papers.
28. Ibid.
29. Ibid., p. 4.
30. Ibid.
31. Ibid., pp. 6–7.
32. Ibid., p. 8.
33. Letter from James W. Watts to Mr. Oliver E. Denham, June 7, 1961. Freeman Papers.
34. Common expression used by inmates in the Philadelphia Prison System, where author Allen Hornblum worked in the 1970s.
35. El-Hai, *The Lobotomist,* p. 267.
36. Howard Dully, *My Lobotomy* (New York: Three Rivers Press, 2007), p. 102.
37. Ibid., p. 91.
38. Quoted in ibid., p. 98.
39. Ibid., p. 268.
40. Shorter, *History of Psychiatry,* p. 227.
41. O. J. Andy, "Thalamotomy in Hyperactive and Aggressive Behavior," *Confinia Neurologica* 32 (1970): 322.
42. Ibid., p. 324.
43. Peter R. Breggin, "Campaigns Against Racist Federal Programs by the Center for the Study of Psychiatry and Psychology," *Journal of African American Men* 1, No. 3 (1995): 6.
44. US Congress, Senate Committee on Labor and Public Welfare, Subcommittee on Health, "Quality of Health Care—Human Experimentation, 1973," Hearings, Ninety-third Congress, first session, on S. 974.
45. Ibid.
46. Benedict Carey, "Surgery for Mental Ills Offers Both Hope and Risk," *New York Times,* November 27, 2009.
47. Braslow, *Mental Ills and Bodily Cures,* p. 143.
48. The word "enthusiast" was repeatedly used in describing the scientific mind-set of Dr. Albert M. Kligman, the famous University of Pennsylvania dermatologist who created lucrative creams for acne and wrinkled skin but also had a long history of using vulnerable populations in his research (See chapter 6). Friends and colleagues of Kligman defended his penchant for sometimes risky, if not dangerous experimentation by explaining "Al was not a venal or menacing person. He was just an enthusiast interested in a wide variety of dermatological subjects." Interviews collected by Allen Hornblum for *Acres of Skin: Human Experiments at Holmesburg Prison: A True Story of Abuse and Exploitation in the Name of Medical Science* (New York: Routledge, 1998).
49. John Marks, *The Search for the Manchurian Candidate* (New York: Norton, 1979), p. 25.

50. Quoted in Martin A. Lee and Bruce Shlain, *Acid Dreams* (Weidenfield, NY: Grove, 1985), p. xxiv.

51. Marks, *Search for the Manchurian Candidate*, p. 28.

52. A number of doctors had questionable histories with LSD research. Harold Abramson, for example, worked for both the army and the CIA and was the last physician (albeit an allergist) to see Frank Olson before he went out the window of his Statler-Hilton hotel room. Paul Hoch was the head of the Psychiatric Institute's mescaline research program that killed Harold Blauer, and Harold Isbell ran the controversial drug research program at Lexington Federal Penitentiary.

53. Lee and Schlain, *Acid Dreams*, p. 45.

54. Lauretta Bender, "Discussion," in *Child Research in Psychopharmacology*, ed. Seymour Fisher (Springfield, IL: Charles C. Thomas, 1959), p. 41.

55. Lee and Schlain, *Acid Dreams*, p. 63.

56. Lauretta Bender, Gloria Faretra, and Leonard Cobrinik, "LSD and UML Treatment of Hospitalized Disturbed Children," *Recent Advances in Biological Psychology* 5 (1963): 37.

57. Ibid., p. 85.

58. Lauretta Bender, "A Twenty-five Year View of Therapeutic Results" *Evaluation of Psychiatric Treatment* (New York: Grune and Stratton, 1964) p. 141.

59. Lauretta Bender, "Children's Reactions to Psychotomimetic Drugs," *Psychotomimetic Drugs* (1970): 268.

60. Ibid., p. 270.

61. Ibid.

## 9 PSYCHOLOGICAL ABUSE

1. Much of our account of the Monster Study comes from Jim Dyer's groundbreaking articles on the study. Jim Dyer, "Ethics and Orphans: The Monster Study," *San Jose Mercury News*, July 25, 2001. According to the *San Jose Mercury News*, Jim Dyer, their reporter on assignment for the newspaper, did not notify authorities of the Iowa State Archive of this fact or of the actual purpose of his research when he approached them for access to the material in question. Although Dyer was a graduate student at the time and claimed to be doing scholarly research, the archive is not open to journalists. The newspaper's editors claimed to be unaware of this ethical breach, and although they viewed Dyer's article as "an important investigative story," they admitted, "It's unfortunate we can't endorse some of the methods used to report it."

2. Quoted in ibid.

3. Ibid.

4. Mary Tudor, "An Experimental Study of the Effect of Evaluative Labelling on Speech Fluency," Master of Arts thesis, Graduate College of the State University of Iowa, August 1939, p. 98.

5. Ibid., p. 67.

6. Quoted in Dyer, "Ethics and Orphans."

7. Tudor, "Experimental Study," pp. 116, 117.

8. Ibid., p. 148.

9. Ibid. p. 147.

10. Dyer, "Ethics and Orphans."

11. Ibid.

12. Quoted in ibid.

13. Tom Owen, "When Words Hurt: Stuttering Study Story Missed the Mark," *The Gazette* (Iowa City), July 12, 2003.

14. Associated Press, "Iowa Pays Victims of Abusive 1930's Stuttering Experiment Paid Almost $1 Million," Fox News, August 20, 2007.

15. Rodney G. Triplet, "Henry A. Murray: The Making of a Psychologist?" *American Psychologist* 47, No. 2 (February 1992): 299–307.

16. Ibid., p. 300.

17. Alston Chase, *Harvard and the Unabomber* (New York: W. W. Norton, 2003), p. 361.

18. Ibid., p. 362.
19. Ibid., p. 363.
20. O. I. Lovaas and James Q. Simmons, "Manipulation of Self-Destruction in Three Retarded Children," *Journal of Applied Behavior Analysis* 2, No. 3 (1969): 149.
21. Ibid., p. 156.
22. Stephanie van Goozen, Walter Matthys, Peggy T. Cohen-Kettenis, Victor Wiegant, and Herman van Engeland, "Salivary Cortisol and Cardiovascular Activity During Stress in Oppositional-Defiant Disorder Boys and Normal Controls," *Biological Psychiatry* 43 (1998): 531.
23. E. Garralda, "Psychophysiological Anomalies in Children with Emotional and Conduct Disorders," *Psychological Medicine* 21 (1991): 947–957.
24. Clemens Kirschbaum, Dirk H. Helhammer, Christian J. Strasburger, Elisabeth Tilling, Renate Kamp, and Harold Luddecke, "Relationships between Salivary Cortisol, Electrodermal Activity and Anxiety under Mild Experimental Stress in Children," *Frontiers of Stress Research* (1989): 383–387.
25. Connie Lamm, I. Granic, P. D. Zelano, and Marc D. Lewis, "Magnitude and Chronometry of Neutral Mechanisms of Emotion Regulation in Subtypes of Aggressive Children," *Brain and Cognition* 77, No. 2 (November 2011): 159–169.

## 10 REPRODUCTION AND SEXUALITY EXPERIMENTS

1. Amy Goodman, "The Case Against Depo-Provera," *Multinational Monitor* 6, Nos. 2 & 3 (February 1985), http://www.multinationalmonitor.org/hyper/issues/1985/02/conference.html.
2. "The Law: Sterilized: Why," *Time Magazine,* July 23, 1973.
3. US Congress, Senate Committee on Labor and Public Welfare, Subcommittee on Health, Quality of Health Care—Human Experimentation, 1973, Hearings, Ninety-third Congress, first session, S. 974. Washington, US Gov't. Printing Office. (Hereafter cited as Human Experimentation hearing.)
4. Authors' phone interviews with Jessie Bly, May 6, 2010, and November 9, 2012.
5. Human Experimentation hearing, p. 74.
6. Ibid., p. 75.
7. Bly interview.
8. Human Experimentation hearing, p. 82.
9. Ibid., p. 77.
10. Ibid., p. 10.
11. Ibid., p. 12.
12. Ibid., p. 120.
13. *Relf vs. Weinberger,* Civil Action No. 1557–73, filed July 31, 1973.
14. Jane Lawrence, "The Indian Health Service and the Sterilization of Native American Women," *American Indian Quarterly* 24, No. 3 (Summer 2000): 400–419.
15. Andrea Smith, *Conquest: Sexual Violence and American Indian Genocide* (Boston: South End Press, 2005), p. 88.
16. Ibid.
17. Elof Axel Carlson, *Times of Triumph, Times of Doubt: Science and the Battle for Public Trust* (New York: Cold Spring Harbor Press, 2006).
18. Everett Flood, "The Advantages of Castration in the Defective," *Journal of the American Medical Association* 29, No. 17 (October 1897): 833.
19. Ibid.
20. Philip Reilly, "The Surgical Solution: The Writings of Activist Physicians in the Early Days of Eugenical Sterilization," *Perspectives in Biology and Medicine* 26 (1983): 646.
21. Marie E. Kopp, "Surgical Treatment as Sex Crime Prevention Measure," *Journal of Criminal Law and Criminology* 28, No. 5 (1938): 693.
22. Louis A. Lurie, "The Endocrine Factor in Homosexuality," *American Journal of the Medical Sciences* 208, No. 2 (August 1944): 180.
23. Harriet Washington, *Medical Apartheid* (New York: Doubleday, 2006), p. 63.

24. Jeffrey S. Sartin, "J. Marion Sims, the Father of Gynecology: Hero or Villain?" *Southern Medical Journal* 97, No. 5 (May 2004): 500.

25. P. Knightley, H. Evans, E. Potter, and M. Wallace, *Suffer the Children: The Story of Thalidomide* (New York: Viking Press, 1979), p. 3.

26. Ibid., p. 45.

27. Ibid., p. 47.

28. Ibid., p. 113.

29. Mary V. Seeman, "Women's Issues in Clinical Trials," in *Clinical Trials in Psychopharmacology: A Better Brain*, ed. Marc Hertzman and Lawrence Adler (Hoboken, NJ: Wiley, 2010), p. 89.

30. A. Goodman, J. Schorge, and M. F. Greene, "The Long-term Effects of In Utero Exposures—The DES Story," *New England Journal of Medicine* 364, No. 22 (2011): 2083–2084.

31. Robert N. Hoover et al., "Adverse Health Outcomes in Women Exposed In Utero to Diethylstilbestrol," *New England Journal of Medicine* 365 (2011): 1304–1314.

32. G. I. M. Swyer, "An Evaluation of the Prophylactic Antenatal Use of Stilboestrol," *American Journal of Obstetrics and Gynecology* 58, No. 5 (1949): 994–1009.

33. Arthur L. Herbst, "Diethistilbestrol and Adenocarcinoma of the Vagina," *American Journal of Obstetrics and Gynecology* 181, No. 6 (1999): 1576–1578.

34. *New York Times* articles from April 16, 1978, and February 27, 1982.

35. Goodman, p. 2084.

36. Eileen Welsome, *The Plutonium Files* (New York: Dial Press, 1999), p. 220.

37. Ibid., p. 221.

38. Ibid., p. 225.

## 11   RESEARCH MISCONDUCT

1. Paul A. Offit, *Autism's False Prophets: Bad Science, Risky Medicine, and the Search for a Cure* (New York: Columbia University Press, 2008), p. 37.

2. Ibid., p. 201.

3. Editorial, "Wakefield's Article Linking MMR Vaccine and Autism was Fraudulent," *British Medical Association* 342 (January 5, 2011).

4. William Broad and Nicholas Wade, *Betrayers of the Truth: Fraud and Deceit in the Halls of Science* (New York: Simon and Schuster, 1982), p. 23.

5. Ibid., p. 37.

6. David J. Miller, "Personality Factors in Scientific Fraud and Misconduct," in *Research Fraud in the Behavioral and Biomedical Sciences*, ed. David J. Miller and Michael Hersen (New York: Wiley, 1992), p. 129.

7. Alexander Kohn, *False Prophets: Fraud and Error in Science and Medicine* (New York: Blackwell, 1986), p. 54.

8. Ibid., p. 55.

9. Broad and Wade, *Betrayers of the Truth*, p. 199.

10. Arthur Jensen, "Scientific Fraud or False Accusations? The Case of Cyril Burt," in Miller and Hersen, *Research Fraud in the Behavioral and Biomedical Sciences*, p. 118.

11. Charles J. Glueck, Margot J. Mellies, Mark Dine, Tammy Perry, and Peter Laskarzewski, "Safety and Efficacy of Long Term Diet and Diet Plus Bile Acid-Binding Resin Cholesterol-Lowering Therapy in 73 Children Heterozygous for Familial Hypercholeserolemia," *Pediatrics* 78, No. 2 (August 1986): 338–348.

12. Mark B. Roman, "When Good Scientists Turn Bad," *Discover* 9, No. 4 (1988): 55, 57.

13. Paul Connett, James Beck, and H. S. Micklem, *The Case against Fluoride: How Hazardous Waste Ended Up in Our Drinking Water and the Bad Science and Powerful Politics That Keep It There* (White River Junction, VT: Chelsea Green, 2010).

14. John Colquhoun, "Why I Changed My Mind about Water Fluoridation," *Perspectives in Biology and Medicine* 41, No. 1 (1997): 29–44.

15. Roman, "When Good Scientists Turn Bad," p. 52.

16. Ibid.

17. Miller, "Personality Factors in Scientific Fraud and Misconduct," p. 136.

18. Alan Poling, "The Consequences of Fraud," in Miller and Hersen, *Research Fraud in the Behavioral and Biomedical Sciences,* p. 146.

## CONCLUSION

1. J. Ipsen, "Bio-assay of Four Tetanus Toxoids in Mice, Guinea Pigs, and Humans," *Journal of Immunology* 70 (1952): 426–434.
2. Interview with Constantine Maletskos, April, 28, 2011, at Spaulding Rehabilitation Hospital, Boston.
3. Jay Katz, *Final Report of the Advisory Committee on Human Radiation Experiments,* (New York: Oxford University Press, 1996), p. 544.
4. In the summer of 1963, Drs. Avir Kagan, Perry Ferskos, and David Leichter refused to participate in a cancer experiment at the Jewish Chronic Disease Hospital and subsequently went to the press to expose what they considered unethical medical research. Subsequent media coverage led to an investigation, several hearings, and two doctors being sanctioned by medical boards.
5. Interview with A. Bernard Ackerman, February 1, 1996.
6. Susan Lederer and Michael Grodin, "Historical Overview: Pediatric Experimentation" in *Children as Research Subjects: Science, Ethics and Law,* ed. Michael A. Grodin and Leonard H. Glantz (New York: Oxford University Press, 1994), pp. 3–25.
7. Ibid.
8. Autonomy suggests that the research participant is a voluntary participant in the research process and has the power to agree to or withdraw from the experiment at any time. The issue of autonomy becomes increasingly complex and dissatisfying when applied to procuring informed consent from underage children and their parents. The principle of justice or fairness concerns equal access to or from research studies and protections against bribes and forms of coercion. The principle of nonmaleficence concerns the assurance that no harm of a physical or psychological nature will result. Veracity means that only the truth about a research project will be communicated, and beneficence means the promotion of good for the participant or for society at large.
9. William Osler, "Experimentation on Man," *Journal of the American Medical Association* 68, No. 5 (February 3, 1917): 373.
10. Donald L. Bartlett and James B. Steele, "Deadly Medicine" *Vanity Fair,* January 2010.
11. Adriana Petryna, *When Experiments Travel: Clinical Trials and the Global Search for Human Subjects* (Princeton, NJ: Princeton University Press, 2009), p. 19.

# BIBLIOGRAPHY

## BOOKS

Annas, George J., and Michael A. Grodin. *The Nazi Doctors and the Nuremberg Code: Human Rights in Human Experimentation*. New York: Oxford University Press, 1992.

Beauchamp, Tom L. *Standing on Principles*. New York: Oxford University Press, 2010.

Benison, Saul. *Tom Rivers: Reflections on a Life in Medicine and Science*. Cambridge, MA: MIT Press, 1967.

Bennett, Abram Elting. *Fifty Years in Neurology and Psychiatry*. New York: Intercontinental Medical Book Corp., 1972.

Birstein, Vadim J. *The Perversion of Knowledge*. New York: Westview Press, 2001.

Black, Edwin. *War against the Weak: Eugenics and America's Campaign to Create a Master Race*. New York: Four Walls Eight Windows, 2003.

Blatt, Burton. *Christmas in Purgatory: A Photographic Essay on Mental Retardation*. Boston: Allyn & Bacon, 1966.

Braslow, Joel. *Mental Ills and Bodily Cures: Psychiatric Treatment in the First Half of the Twentieth Century*. Berkeley: University of California Press, 1997.

Breggin, Peter, and Ginger Breggin. *The War against Children of Color: Psychiatry Targets Inner City Youth*. Monroe, ME: Common Courage Press, 1998.

Bruinius, Harry. *Better for All the World: The Secret History of Forced Sterilization and America's Quest for Racial Purity*. New York: Knopf, 2006.

Buchanan, David, Celia Fisher, and Lance Gable, eds. *Research with High-Risk Populations: Balancing Science, Ethics, and Law*. Washington, DC: American Psychological Association, 2009.

Carlson, Elof Axel. *Times of Triumph, Times of Doubt: Science and the Battle for Public Trust*. Cold Spring Harbor, NY: Cold Spring Harbor Laboratory Press, 2006.

Chase, Alston. *Harvard and the Unabomber*. New York: W. W. Norton, 2003.

Cole, Leonard A. *The Eleventh Plague*. New York: W. H. Freeman, 1997.

Collins, Anne. *In the Sleep Room: The Story of the CIA Brainwashing Experiments in Canada*. Toronto: Lester & Orpen Dennys, 1988.

D'Antonio, Michael. *The State Boys Rebellion: The Inspiring True Story of American Eugenics and the Men Who Overcame It*. New York: Simon & Schuster, 2004.

Davis, Guy Pratt. *What Shall the Public Schools Do for the Feeble-Minded? A Plan for Special-School Training under Public School Auspices*. Cambridge, MA: Harvard University Press, 1927.

De Kruif, Paul. *Life Among the Doctors*. New York: Harcourt, Brace, 1949.

De Kruif, Paul. *Microbe Hunters*. New York: Harcourt, Brace, 1926.

De Kruif, Paul. *The Sweeping Wind*. New York: Harcourt, Brace, & World, 1962.

Dorr, Gregory Michael. *Segregation's Science: Eugenics and Society in Virginia*. Charlottesville: University of Virginia Press, 2008.

Dowdall, George W. *The Eclipse of the State Mental Hospital: Policy, Stigma, and Organization.* Albany: State University of New York Press, 1996.

DuBois, James M., Jill E. Ciesla, and Kevin E. Voss. "Research Ethics in US Medical Education: An Analysis of Ethics Course Syllabi." In *Research of Research Integrity*, ed. by N. H. Steneck and M. D. Sheetz. Bethesda, MD: Department of Health and Human Services, 2001.

Dully, Howard, and Charles Fleming. *My Lobotomy.* New York: Three Rivers Press, 2007.

El-Hai, Jack. *The Lobotomist: A Maverick Medical Genius and His Tragic Quest to Rid the World of Mental Illness.* Hoboken, NJ: John Wiley & Sons, 2005.

Epstein, Steven. *Inclusion: The Politics of Difference in Medical Research.* Chicago: University of Chicago Press, 2007.

Flexner, Abraham. *Abraham Flexner: An Autobiography.* New York: Simon & Schuster, 1960.

Goliszek, Andrew. *In the Name of Science.* New York: St. Martin's Press, 2003.

Goodman, Jordan, Anthony McElligott, and Lara Marks. *Useful Bodies: Humans in the Service of Medical Science in the Twentieth Century.* Baltimore: Johns Hopkins University Press, 2003.

Guerrini, Anita. *Experimenting with Humans and Animals: From Galen to Animal Rights.* Baltimore: Johns Hopkins University Press, 2003.

Harris, Sheldon H. *Factories of Death: Japanese Biological Warfare, 1932–45 and the American Cover-up.* New York: Routledge, 1994.

Harvey, A. McGehee. *Science at the Bedside: Clinical Research in American Medicine, 1905–1945.* Baltimore: Johns Hopkins University Press, 1981.

Heberer, Patricia, and Jurgen Matthaus, eds. *Atrocities on Trial: Historical Perspectives on the Politics of Prosecuting War Crimes.* Lincoln: University of Nebraska Press, 2008.

Hooper, Edward. *The River: A Journey to the Source of HIV and AIDS.* New York: Little, Brown, 1999.

Hoover, J. Edgar. *Masters of Deceit: The Story of Communism in America and How to Fight It.* New York: Henry Holt, 1958.

Hornblum, Allen M. *Acres of Skin: Human Experiments at Holmesburg Prison; A True Story of Abuse and Exploitation in the Name of Medical Science.* New York: Routledge, 1998.

Institute of Medicine. *Ethical Conduct of Clinical Research Involving Children.* Washington, DC: National Academy of Sciences, 2004.

Ketchum, James S. *Chemical Warfare: Secrets Almost Forgotten.* Self-published, 2006.

Kevles, Daniel J. *In the Name of Eugenics: Genetics and the Uses of Human Heredity.* New York: Knopf, 1985.

Knightley, P., H. Evans, E. Potter, and M. Wallace, *Suffer the Children: The Story of Thalidomide* (New York: Viking Press, 1979).

Kraut, Alan M. *Goldberger's War: The Life and Work of a Public Health Official.* New York: Hill and Wang, 2003.

Lederer, Susan E. *Subjected to Science: Human Experimentation in America before the Second World War.* Baltimore: Johns Hopkins University Press, 1995.

Lewis, Sinclair. *Arrowsmith.* New York: Signet Classic, 1998.

Moreno, Jonathan. *Undue Risk: Secret State Experiments on Humans.* New York: W. H. Freeman, 2000.

Noll, Steven. *Feeble-Minded in Our Midst: Institutions for the Mentally Retarded in the South, 1900–1940.* Chapel Hill: University of North Carolina Press, 1995.

Noll, Steven, and James W. Trent Jr. *Mental Retardation in America.* New York: New York University Press, 2004.

Oldstone, Michael B. A. *Viruses, Plagues, & History.* New York: Oxford University Press, 2010.

Oshinsky, David M. *Polio: An American Story.* New York: Oxford University Press, 2005.

Pappworth, M. H. *Human Guinea Pigs: Experimentation on Man.* Boston: Beacon Press, 1967.

Parascandola, John. *The Development of American Pharmacology.* Baltimore: Johns Hopkins University Press, 1992.

Parsons, Robert P. *Trail to Light: A Biography of Joseph Goldberger.* New York: Bobbs-Merrill, 1943.

Petryna, Adriana. *When Experiments Travel: Clinical Trials and the Global Search for Human Subjects.* Princeton, NJ: Princeton University Press, 2009.

Poling, Alan. "The Consequences of Fraud." In *Research Fraud in the Behavioral and Biomedical Sciences*, ed. by David Miller and Michael Hersen. New York: John Wiley & Sons, 1992.

Rhodes, Richard. *Dark Sun: The Making of the Hydrogen Bomb.* New York: Simon & Schuster, 1995.

Rosenberg, Charles E. *The Care of Strangers: The Rise of America's Hospital System.* New York: Basic Books, 1987.

Rosenberg, Charles E. *No Other Gods: On Social Thought.* Baltimore: Johns Hopkins University Press, 1997.

Ross, Colin. *Bluebird: Deliberate Creation of Multiple Personality by Psychiatrists.* Richardson, TX: Manitou Communications, 2000.

Ross, Lainie Friedman. *Children in Medical Research: Access versus Protection.* New York: Oxford University Press, 2006.

Rothman, David J. *Strangers at the Bedside.* New York: Basic Books, 1991.

Schmidt, Ulf. *Justice at Nuremberg: Leo Alexander and the Nazis Doctors' Trial.* New York: Palgrave, 2006.

Scull, Andrew. *Madhouse: A Tragic Tale of Megalomania and Modern Medicine.* New Haven, CT: Yale University Press, 2005.

Shorter, Edward, and David Healy. *Shock Therapy: A History of Electroconvulsive Treatment in Mental Illness.* New Brunswick, NJ: Rutgers University Press, 2007.

Simpson, Christopher. *Blowback: The First Full Account of America's Recruitment of Nazis, and Its Disastrous Effect on Our Domestic and Foreign Policy.* New York: Weidenfeld & Nicolson, 1988.

Skloot, Rebecca. *The Immortal Life of Henrietta Lacks.* New York: Crown, 2010.

Slaughter, Frank G. *Daybreak: The Story of a Young Doctor Faced with a Momentous Challenge.* New York: Doubleday, 1958.

Slaughter, Frank G. *East Side General: A Violent Drama of Death and Love in a Big City.* New York: Permabook, 1952.

Slaughter, Frank G. *Spencer Brade, M.D.* New York: Doubleday, 1942.

Spitz, Vivien. *Doctors from Hell: The Horrific Account of Nazi Experiments on Humans.* Boulder, CO: Sentient, 2005.

Starr, Paul. *The Social Transformation of American Medicine.* New York: Basic Books, 1982.

Stern, Alexandra Minna. *Eugenic Nation: Faults and Frontiers of Better Breeding in Modern America.* Berkeley: University of California Press, 2005.

Tierney, Patrick. *Darkness in El Dorado: How Scientists and Journalists Devastated the Amazon.* New York: W. W. Norton, 2000.

Trent, James W. Jr. *Inventing the Feeble Mind: A History of Mental Retardation in America.* Berkeley: University of California Press, 1994.

Tyor, Peter L., and Leland V. Bell. *Caring for the Retarded in America: A History.* Westport, CT: Greenwood Press, 1984.

Vaughan, Roger. *Listen to the Music: The Life of Hilary Kroprowski.* New York: Springer, 2000.

Weinstein, Allen, and Alexander Vassiliev. *The Haunted Wood: Soviet Espionage in America.* New York: Random House, 1999.

Weinstein, Harvey. *Father, Son and the CIA.* Halifax, Nova Scotia: Goodread Biographies, 1990.

Welsome, Eileen. *The Plutonium Files: America's Secret Medical Experiments in the Cold War.* New York: Dial Press, 1999.

Whitaker, Robert. *Mad in America: Bad Science, Bad Medicine, and the Enduring Mistreatment of the Mentally Ill.* New York: Perseus, 2002.

Whitney, Leon F. *The Case for Sterilisation.* London: John Lane, 1935.

Zenderland, Leila. *Measuring Minds: Henry Herbert Goddard and the Origins of American Intelligence Testing.* New York: Cambridge University Press, 1998.

Zimbardo, Philip. *The Lucifer Effect: Understanding How Good People Turn Evil.* New York: Random House, 2007.

## ARTICLES

Abt, Isaac A. "Spontaneous Hemorrhages in Newborn Children." *Journal of the American Medical Association* 40, No. 5 (January 1903): 200.

Allen, Scott. "MIT Official Tells More of Radiation Experiments at Fernald," *Boston Globe*, December 30, 1993.

Allen, Scott. "MIT Records Show Wider Radioactive Testing at Fernald," *Boston Globe*, December 31, 1993.

Allen, Scott. "Radiation Used on Retarded," *Boston Globe*, December 26, 1993.

Altman, Lawrence. "Immunization Is Reported in Serum Hepatitis Tests," *New York Times*, March 24, 1971.

Andy, O. J. "Neurosurgical Treatment of Abnormal Behavior." *American Journal of the Medical Sciences* 252, No. 2 (1966): 232–238.

Andy, O. J. "Thalamotomy in Hyperactive and Aggressive Behavior." *Confinia Neurologica* 32 (1970): 332.

Armstrong, David. "Experiment Records Search Ordered at State Facilities," *Boston Globe*, December 31, 1993.

Armstrong, David. "State Expects Further 1940s–50s Revelations," *Boston Globe*, December 29, 1993.

Associated Press. "Iowa Pays Victims of Abusive 1930's Stuttering Experiment Paid Almost $1 Million," Fox News, August 20, 2007.

Barker, Lewellys. "The Importance of the Eugenic Movement and Its Relation to Social Hygiene." *Journal of the American Medical Association* 54 (1910): 2017–2022.

Barnes, Donald. "Effect of Evaporated Milk on the Incidence of Rickets in Infants." *Journal of the Michigan State Medical Society* 31 (1932): 397.

Barr, Martin W. "Some Notes on Asexualization with a Report on 18 Cases." *Journal of Nervous & Mental Disease* 51, No. 3 (March 1920): 232.

Bartlett, Donald L., and James B. Steele. "Deadly Medicine," *Vanity Fair*, January 2010.

Beecher, Henry K. "Ethics and Clinical Research." *New England Journal of Medicine* 274, No. 24 (June 16, 1966): 1354–1358.

Benda, C. E., C. J. Maletskos, J. C. Hutchinson, and E. B. Thomas. "Studies of Thyroid Function in *Myotonia Dysttrophien*." *American Journal of Medical Sciences* 228 (1954): 668–672.

Bender, Lauretta. "Children's Reactions to Psychotomimetic Drugs." *Psychotomimetic Drugs* (1970): 268.

Bender, Lauretta. "D-Lysergic Acid in the Treatment of the Biological Features of Childhood Schizophrenia." *Diseases of the Nervous System* 27 (1966): 43–46.

Bender, Lauretta. "One Hundred Cases of Childhood Schizophrenia Treated with Electric Shock." *Transactions of the American Neurologic Association* 762 (1947): 168.

Bender, Lauretta. "Theory and Treatment of Childhood Schizophrenia." *Acta Paedopsychiatrica* 34 (1968): 301.

Bender, Lauretta, Gloria Faretra, and Leonard Gobrinik. "LSD and UML Treatment of Hospitalized Disturbed Children." *Recent Advances in Biological Psychology* 5 (1963): 37.

Bennett, A. E. "Preventing Traumatic Complications in Convulsive Shock Therapy by Curare." *Journal of the American Medical Association* 114, No. 4 (1940): 322–324.

Bibby, Basil G. "A Test of the Effect of Fluoride-Containing Dentifrices." *Journal of Dental Research* 24, No. 6 (1945): 297–303.

Black, W. C. "The Etiology of Acute Infectious Gingivostomatitis." *Journal of Pediatrics* 30 (1942): 145–160.

Boyd, Julian D., and Kenneth E. Wessels. "Epidemiologic Studies in Dental Caries." *American Journal of Public Health* 41 (August 1951): 978.

Bradley, Charles. "The Behavior of Children Receiving Benzedrine." *American Journal of Psychiatry* 94 (1937): 577–585.

Brave, Ralph, and Kathryn Sylva. "Exhibiting Eugenics Response and Resistance to a Hidden History." *Public Historian* 29, No. 3 (Summer 2007): 37.

Breggin, Peter R. "Campaigns Against Racist Federal Programs by the Center for the Study of Psychiatry and Psychology." *Journal of African American Men* 1, No. 3 (1995): 6.

Brodbeck, Arthur J., and Orvis C. Irwin. "The Speech Behavior of Infants Without Families." *Child Development* 17, No. 3 (September 1946): 145–156.

Brunner, Felix, Robert S. Harris, Constantine J. Maletskos, and Clemens E. Benda. "Studies in Calcium Metabolism. Effects of Food Phytates on Calcium Uptake in Children on Low-Calcium Breakfasts." *Journal of Nutrition* 54, No. 4 (December 1954): 523–542.

Buckman, John. "Lysergic Acid Diethylamide." *British Medical Journal*, July 30, 1966, p. 302.

Buckman, John. "Theoretical Aspects of L.S.D. Therapy." *International Journal of Social Psychiatry* 13, No. 2 (1967): 126–138.

Burman, Michael S. "Therapeutic Use of Curare and Erythroidine Hydrochloride for Spastic and Dystonic States." *Archives of Neurology and Psychiatry* 41 (1939): 307–327.

Campbell, Eric G. "Financial Relationships Between Institutional Review Board Members and Industry." *New England Journal of Medicine* 355, No. 22 (November 2006): 2321–2329.

Carey, Benedict. "Surgery for Mental Ills Offers Both Hope and Risk," *New York Times,* November 27, 2009.

Collins, Tom, and Selwyn Raab. "Charge Hospital Shot Live Cancer Cells into Patients," *New York World-Telegram,* January 20, 1964.

Colquhoun, John. "Why I Changed My Mind about Water Fluoridation." *Perspectives in Biology and Medicine* 41, No. 1 (Autumn 1997): 29–44.

Coombs, Francis, P., and Thomas Butterworth. "Atypical Keratosis Pilaris." *Archives of Dermatology* 62, No. 2 (August 1950): 305.

Curran, William J., and Henry K. Beecher. "Experimentation in Children." *Journal of the American Medical Association* 10, No. 1 (October 6, 1969): 77–83.

Daniel, F. E., and S. E. Hudson. "Emasculation of Masturbation—Is It Justifiable?" *Texas Medical Journal* 10, No. 4 (1895): 239.

Darby, William, Paul Hahn, Margaret Kaser, Ruth Steinkamp, Paul Deneen, and Mary Cook. "The Absorption of Radioactive Iron by Children 7–10 Years of Age." *Journal of Nutrition* 33, No. 1 (1947): 107–119.

Dowling, G. B. "The Treatment of Ringworm of the Scalp by Thallium Depilation." *British Medical Journal* 2, No. 3475 (August 1927): 261–263.

"Dr. Brodie Upholds Paralysis Vaccine," *New York Times,* November 3, 1935.

Dyer, Jim. "Ethics and Orphans: The Monster Study," *San Jose Mercury News,* July 25, 2001.

Editorial. "Effects of L.S.D." *British Medical Journal,* No. 5502, June 18, 1966, p. 1495.

Editorial. "Fernald's Atomic Café," *Boston Globe,* December 29, 1993.

Editorial. "Lobotomy Disappointment," *Newsweek,* December 12, 1949, p. 51.

Editorial. "The Nuremberg Trial against German Physicians." *Journal of the American Medical Association* 135, No. 13 (November 1947): 135.

Editorial. "Poliomyelitis: A New Approach." *Lancet* (March 1952): 552.

Editorial. "Therapeutic Orphans." *Journal of Pediatrics* 72, No. 1 (January 1968): 119–120.

Editorial. "The Treatment of Diphtheria Carriers." *Journal of the American Medical Association* 67, No. 19 (November 4, 1916): 1372.

Editorial. "Wakefield's Article Linking MMR Vaccine and Autism Was Fraudulent." *British Medical Journal* 342 (January 2011): C7452.

Elliot, Carl. "Guinea-Pigging," *The New Yorker,* January 7, 2008, pp. 36–41.

Elmore, Joann G., and Alvan Feinstein. "Joseph Goldberger: An Unsung Hero of American Clinical Epidemiology." *Annals of Internal Medicine* 121, No. 5 (September 1994): 372.

Engler, M. "Prefrontal Leucotomy in Mental Defectives." *Journal of Mental Science* 94 (October 1948): 844–850.

Falusi, Adeyinka, O. I. Olopade, and C. O. Olopade. "Establishment of a Standing Ethics/Institutional Board in a Nigerian University." *Journal of Empirical Research on Human Research Ethics* 2, No. 1 (March 2007): 21–30.

Felden, Botha. "Epilation with Thallium Acetate in the Treatment of Ringworm of the Scalp of Children." *Archives of Dermatology and Syphilology* 17, No. 2 (February 1928): 185.

Finn, S. B., and H. C. Jamison. "The Effect of a Dicalcium Phosphate Chewing Gum on Caries Incidence in Children." *Journal of the American Medical Association* 74, No. 5 (April 1967): 987–995.

Flood, Everett. "Notes on the Castration of Idiot Children." *American Journal of Psychology* 10, No. 2 (January 1899): 300.

Flood, Everett. "The Advantages of Castration in the Defective." *Journal of the American Medical Association* 29, No. 17 (October 1897): 833.

Frederick, Delores. "Experiments on Humans Denied Here." *Pittsburgh Press,* April 12, 1973.

Frederick, Delores. "Courts, Not Parents, Have Word on Retarded Tests, Wecht Says," *Pittsburgh Press,* April 12, 1973.

Freeman, Walter. "Psychosurgery: Present Indications and Future Prospects." *Journal of the California Medical Association* 88, No. 6 (June 1958): 429–434.

Galvin, Kevin. "Kennedy Plans Hearings on Radiation Tests," *Boston Globe,* December 30, 1993.

Geison, Gerald L. "Pasteur's Work on Rabies: Reexamining the Ethical Issues." *Hastings Center Report* 8 (1978): 26–33.

Glantz, Leonard H. "Protecting Children and Society." *Journal of Medical Ethics* 7, No. 2 (June 1979): 4–5.

Glass, R. L. "Effects on Denial Caries Incidence of Frequent Ingestion of Small Amounts of Sugars and Stannous EDTA from Chewing Gum." *British Dental Journal* 145 (1981): 95–100.

Glueck, Charles J., Margot J. Mellies, Mark Dine, Tammy Perry, and Peter Laskarzewski. "Safety and Efficacy of Long Term Diet and Diet Plus Bile Acid-Binding Resin Cholesterol-lowering Therapy in 73 Children Heterozygous for Familial Hypercholestarolemia." *Pediatrics* 78, No. 2 (August 1986): 338–348.

Goldberger, Joseph. "The Etiology of Pellagra." *Public Health Reports* 29, No. 26 (June 26, 1914): 373–375.

Goldberger, Joseph. "Experimental Pellagra in the Human Subject Brought About by a Restricted Diet." *Public Health Reports* 30, No. 46 (November 12, 1915): 3336–3339.

Goldman, Herbert, Samuel, Karelitz, and Norman Schell. "Acidosis in Premature Infants Due to Lactic Acid." *Pediatrics* 27 (1961): 928.

Goldschmidt, Herbert, and Albert Kligman. "Experimental Inoculation of Humans with Ectodermotropic Viruses." *Journal of Investigative Dermatology* 31 No. 3 (September 1958): 176.

Gorski, David. "On the Ethics of Clinical Trials of Homeopathy in Third World Countries." *Science Based Medicine* (March 2008), http://www.sciencebasedmedicine.org.

Hamill, Samuel M., Howard C. Carpenter, and Thomas A. Cope. "A Comparison of the Pirquet, Calmette, and Moro Tuberculin Tests and Their Diagnostic Value." *Archives of Internal Medicine* 2 (1908): 405.

Hansen, Bert. "Medical History for the Masses: How American Comic Books Celebrated Heroes of Medicine in the 1940s." *Bulletin of the History of Medicine* 78, No. 1 (Spring 2004): 148–191.

Harcourt, Bernard E. "Making Willing Bodies: Manufacturing Consent Among Prisoners and Soldiers, Creating Human Subjects, Patriots, and Everyday Citizens—The University of Chicago Malaria Experiments on Prisoners at Stateville Penitentiary." February 6, 2011. University of Chicago Law & Economics, Olin Working Paper No. 544; University of Chicago, Public Law Working Paper No. 341. Available at SSRN: http://ssrn.com/abstract=1758829 or http://dx.doi.org/10.2139/ssrn.1758829.

Henle, Werner, Gertrude Henle, and Joseph Stokes Jr. "Demonstration of the Efficacy of Vaccination Against Influenza Type A by Experimental Infection of Human Beings." *Journal of Immunology* 46 (1943): 163–175.

Holt, L. Emmett. "A Report Upon One Thousand Tuberculin Tests in Young Children." *Archives of Pediatrics* (January 1909): 1–10.

Hoover, Robert N. "Adverse Health Outcomes in Women Exposed in Utero to Diethylstilbestrol." *New England Journal of Medicine* 365 (2011): 1304–1314.

Howe, Howard A. "Antibody Response of Chimpanzee and Human Beings to Formalin Inactivated Trivalent Poliomyelitis Vaccine." *Journal of Experimental Medicine* 80, No. 5 (November 1944): 383–390.

Howe, Howard A. "Criteria for the Inactivation of Poliomyelitis." Committee on Immunization, December 4, 1951 Meeting, Mandeville Special Collections Library, University of California, San Diego, 3.

Hudson, N., Edwin Lennette, and Francis H. Gordon. "Factors of Resistance in Experimental Poliomyelitis." *Journal of the American Medical Association* 106, No. 24 (June 1936): 1.

Ipsen, J. "Bio-assay of Four Tetanus Toxoids in Mice, Guinea Pigs, and Humans." *Journal of Immunology* 70 (1952): 426–434.

Ireton, Gabriel. "Polk Chaplin Defends Use of Patient Pens," *Pittsburgh Post-Gazette,* June 21, 1973.

Jervis, George A., Ferdinand F. McAllister, and Bruce M. Hogg. "Revascularization of the Brain in Mental Defectives." *Neurology* 3 (1953): 871–878.

Kaempffert, Waldemar. "Turning the Mind Inside Out," *Saturday Evening Post,* May 24, 1941.

Kligman, Albert M. "Hyposensitization Against Rhus Dermatitis." *Archives of Dermatology* 78 (July 1958): 47–72.

Kligman, Albert M. "The Pathogenesis of Tinea Capitis Due to Microsporum Audouini and Microsporum Canis." *Journal of Investigative Dermatology* 18, No. 3 (March 1952): 231.

Kligman, Albert M., and W. Ward Anderson. "Evaluation of Current Methods for the Local Treatment of Tinea Capitis." *Journal of Investigative Dermatology,* No. 16 (March 1954): 162.

Knowles, Frank Crozer. "*Molluscum Contagiosum:* Report of an Institutional Epidemic of Fifty-Nine Cases." *Journal of the American Medical Association* 53, No. 9 (August 28, 1909): 671.

Knox, Richard. "Studies Fail by Today's Standards," *Boston Globe,* December 31, 1993.

Kong, Delores. "Ex-Resident Recalls Fernald Club," *Boston Globe,* December 29, 1993.

Koprowski, Hilary, Thomas W. Norton, Klaus Hummeler, Joseph Stokes Jr., Andrew D. Hunt, Agnes Flack, and George A. Jervis. "Immunization of Infants with Living Attenuated Poliomyelitis Virus." *Journal of the American Medical Association* 162, No. 14 (December 1, 1956): 1281–1288.

Krugman, Saul. "The Willowbrook Hepatitis Studies Revisited: Ethical Aspects." *Reviews of Infectious Diseases* 8, No. 1 (January 1986): 158.

Krugman, Saul, Robert Ward, Joan Giles, Oscar Bodansky, and A. Milton Jacobs. "Infectious Hepatitis: Detection of Virus during the Incubation Period and in Clinically Inapparent Infection." *New England Journal of Medicine* 261, No. 15 (October 8, 1959): 729–734.

Kurland, J. G., S. Fishman, M. W. Hamolsky, and A. S. Freedberg. "Radioisotope Study of Thyroid Function in 21 Mongoloid Subjects Including Observations in 7 Parents." *Journal of Clinical Endochronology and Metabolism* 17 (1957): 552–560.

Ladimer, Irving. "Ethical and Legal Aspects of Medical Research on Human Beings." *Journal of Public Law* 3 (1954): 467–511.

Lamm, Connie, L. Granie, P. D. Zelano, and Marc D. Lewis. "Magnitude and Chronometry of Neutral Mechanisms of Emotion Regulation in Subtypes of Aggressive Children." *Brain and Cognition* 77, No. 2 (November 2011): 159–169.

Lawrence, Jane. "The Indian Health Service and the Sterilization of Native American Women." *American Indian Quarterly* 24, No. 3 (Summer 2000): 400–419.

Lederer, Susan. "Hideyo Noguchi's Lutein Experiment and the Antivivisectionists." *Isis* 76 (1985): 31–48.

Lederer, Susan. "Political Animals: The Shaping of Biomedical Research Literature in Twentieth-Century America." *Isis* 83 (1992): 61–79.

Lee, Gary. "U.S. Should Pay Victims, O'Leary Says," *Washington Post,* December 29, 1993.

Lippmann, Walter. "The Abuse of the Tests," *New Republic,* November 15, 1922, p. 297.

Lombardo, Paul A. "Tracking Chromosomes, Castrating Dwarves: Uninformed Consent and Eugenic Research." *Ethics & Medicine* 25, No. 3 (Fall 2009): 149–164.

Lombardo, Paul, A., and Gregory Dorr. "Eugenics, Medical Education and the Public Health Service: Another Perspective on the Tuskegee Syphilis Experiment." *Bulletin of the History of Medicine* 80, No. 2 (Summer 2006): 302.

Lovaas, O. I., and James Q. Simmons. "Manipulation of Self-Destruction in Three Retarded Children." *Journal of Applied Behavior Analysis* 2, No. 3 (1969): 149.

Lowery, George H., Walter H. Beierwalters, Isadore Lampe, and Henry Gomberg. "Radioiodine Uptake Curve in Humans." *Pediatrics* 4 (1949): 627–633.

Lurie, Louis A. "The Endocrine Factor in Homosexuality." *American Journal of the Medical Sciences* 208, No. 2 (August 1944): 180.

Lydston, G. Frank. "Sex Mutilations in Social Therapeutics." *New York Medical Journal* 95 (1912): 677–685.

Mack, Pauline B., Alice Knapper Milsom, and Paul L. Carney. "A Study of Two Levels of Bread Enrichment in Children's Diet." *Monographs of the Society for Research in Child Development* 18, No. 2 (1953): 1–92.

Maris, Elizabeth P., Geoffrey Rake, Joseph Stokes Jr., Morris F. Shaffer, and Gerald C. O'Neil. "Studies on Measles: The Results of Chance and Planned Exposure to Unmodified Measles Virus in Children Previously Inoculated with Egg Passage Measles Virus." *Journal of Pediatrics* 22, No. 1 (January 1943): 17.

Martmer, Edgar E., Kenneth E. Corrigang, Harold Chareneau, and Allen Sosin. "A Study of the Uptake of Iodine by the Thyroid of Premature Infants." *Pediatrics* 17 (1955): 507.

Mashek, John. "Official Asks Damages for Test Victims," *Boston Globe,* December 29, 1993.

Matsuyama, Steven S., and Lissy F. Jarvik. "Cytogenetic Effects of Psychoactive Drugs." *Genetics and Pharmacology* 10 (1975): 99–132.

Mazur, Allan. "Allegations of Dishonesty in Research and Their Treatment by American Universities." *Minerva* 27 (1989): 177.

McChamill, Samuel, Howard Childs Carpenter, and Thomas Cope. "A Comparison of the Von Pirquet, Calmette, and Moro Tuberculin Tests and Their Diagnostic Value." *Archives of Internal Medicine* 2, No. 5 (1908): 405–447.

Middlesworth, Van. "Radioactive Iodide Uptake of Normal Newborn Infants." *American Journal of Diseases in Children* 38 (1954): 439.

Millan, Stephan, John Maisel, Henry Kempe, Stanley Plotkin, Joseph Pagano, and Joel Warren. "Antibody Response of Man to Canine Distemper Virus." *Journal of Bacteriology* 79, No. 4 (1960): 618.

Mitchell, Ross G. "The Child and Experimental Medicine." *British Medical Journal,* March 21, 1964, p. 721.

Najjar, Victor, and L. Emmett Holt, "The Biosynthesis of Thiamine in Human Nutrition." *Journal of the American Medical Association* 123, No. 11 (November 1943): 683–684.

Najjar, Victor, George Johns, George Mediary, Gertrude Fleischmann, and L. Emmett Holt. "The Biosynthesis of Riboflavins." *Journal of the American Medical Association* 126, No. 6 (October 1944): 357–358.

Noguchi, Hideyo. "Experimental Research in Syphilis." *Journal of the American Medical Association* 58, No. 16 (April 20, 1912): 1163–1171.

"New Infantile Paralysis Vaccine Is Declared to Immunize Children," *New York Times,* August 18, 1934.

Ogborn, Richard E., Ronald E. Waggener, and Eugene VanHove. "Radioactive Iodine Concentration in Thyroid Glands of Newborn Infants." *Pediatrics* 28 (1960): 771.

Owen, Tom. "When Words Hurt: Stuttering Study Story Missed the Mark," *Gazette* (Cedar Rapids, Iowa), July 12, 2003.

"Paralysis Vaccine Discontinued Here," *New York Times,* December 27, 1935.

Pierce, Henry W. "State Halts Use of Retarded as Guinea Pigs," *Pittsburgh Post-Gazette,* April 11, 1973.

Pierce, Henry W. "State Kills Testing of Meningitis Shots," *Pittsburgh Post-Gazette,* April 12, 1973.

Pisek, Godfrey R., and Leon Theodore LeWald. "The Further Study of Immunization with Living Rubella Virus Trials in Children with a Strain Cultured from an Aborted Fetus." *American Journal of Disability and Children* 110 (October 1965): 382.

Plotkin, Stanley, David Cornfield, and Theodore H. Ingalls. "Studies of Immunization with Living Rubella Virus Trials in Children with a Strain Cultured from an Aborted Fetus." *American Journal of Disability and Children* 110 (October 1965): 382.

Plotkin, Stanley, J. D. Farquhar, M. Katz, and C. Hertz. "Further Studies of an Attenuated Rubella Strain Grown in WI-38 Cells." *American Journal of Epidemiology* 89, No. 2 (September 6, 1968): 238.

Porcaro, Edward T. "Experimentation with Children." *Journal of Law and Medical Ethics* 7, No. 2 (June 1979): 7–9.

"Prison Malaria: Convicts Expose Themselves to Disease So Doctors Can Study It," *Life,* June 4, 1945, 43-46.

"Psychosurgery," *Time,* November 30, 1942.

Puga, Anna. "Energy Chief Fighting a Culture of Secrecy," *Boston Globe,* December 31, 1993.

"Questions Safety of Paralysis Virus," *New York Times,* October 9, 1935.

Reilly, Philip. "The Surgical Solution: The Writings of Activist Physicians in the Early Days of Eugenical Sterilization." *Perspectives in Biology and Medicine* 26 (1983): 646.

Rezendes, Michael. "For Markey, Belated Vindication," *Boston Globe,* December 30, 1993.

Rezendes, Michael. "Rep. Markey Vindicated by U.S. Disclosure of Radiation Experiments," *Washington Post,* December 31, 1993.

Ritts, R. E. "A Physician's View of Informed Consent in Human Experimentation." *Fordham Law Review* 36 (1967–68): 631–638.

Roman, Mark B. "When Good Scientists Turn Bad." *Discover* 9, No. 4 (1988): 55.

Rothman, David J. "Ethics and Human Experimentation." *New England Journal of Medicine* 317, No. 19 (November 5, 1987): 1195–1199.

Rothman, David J. "Were Tuskegee and Willowbrook 'Studies in Nature'?" *Hastings Center Report* 12, No. 2 (April 1982): 5–7.

Rowe, N. H., R. H. Anderson, and L. A. Wanninger. "Effects of Ready to Eat Breakfast Cereals on Dental Caries Experience in Adolescent Children." *Journal of Dental Research* 53, No. 1 (January 1974): 33–36.

Salk, Jonas E., Harold E. Pearson, Philip N. Brown, and Thomas Francis Jr. "Protective Effect of Vaccination Against Influenza B." *Journal of Clinical Investigation* 24 (July 1945): 547–553.

Saltus, Richard. "Some Fernald Research Defended," *Boston Globe,* December 28, 1993.

Sartain, Jeffrey S. "J. Marion Sims, the Father of Gynecology: Hero or Villain?" *Southern Medical Journal* 97, No. 5 (May 2004): 500.

Sauer, Louis W., and Winston H. Tucker. "Immune Responses to Diphtheria, Tetanus, and Pertussis, Aluminum Phosphate Adsorbed." *Journal of Public Health* 44, No. 6 (1954): 785.

Saxena, Krishna M., Earle M. Chapman, and Charles V. Pryles. "Minimal Dosage of Iodide Required to Suppress Uptake of Iodide-131 by Normal Thyroid." *Science* 138 (1962): 430.

Saxena, Krishna M., and C. V. Pryles. "Thyroid Function in Mongolism." *Journal of Pediatrics* 67 (1965): 363–370.

Schaeffer, Morris. "William H. Park: His Laboratory and His Legacy." *American Journal of Public Health* 75, No. 11 (November 1985): 1301.

Schneider, Keith. "1950 Memo Shows Radiation Test Doubts." *New York Times,* December 28, 1993.

Schneider, Keith. "Energy Official Seeks to Assist Victims of Tests." *New York Times,* December 29, 1993.

Schneider, Keith. "U.S. Expands Inquiry into Its Human Radiation Tests." *New York Times,* December 31, 1993.

Seidelman, William E. "Mengele Medicus: Medicine's Nazi Heritage." *Milbank Quarterly* 66, No. 2 (1988): 221–235.

Sharpe, Leon M., Wendell C. Peacock, Richard Cooke, and Robert S. Harris. "The Effect of Phytate and Other Food Factors on Iron Absorption." *Journal of Nutrition* (1949): 433–445.

Shribman, David. "Clinton Orders Conference on Radiation Tests." *Boston Globe,* December 31, 1993.

Shribman, David. "Stepping Back from Progress," *Boston Globe,* December 31, 1993.

Steinberg, A. D., S. O. Zimmerman, and M. L. Bramer. "The Lincoln Dental Caries Study: The Effect of Acidulated Carbonated Beverages on the Increase of Dental Caries." *Journal of the American Dental Association* 85, No. 1 (1972): 81–89.

Sternberg, George M., and Walter Reed. "Report on Immunity against Vaccination Conferred upon the Monkey by the Use of the Serum of the Vaccinated Calf and Monkey." *Transactions of the Association of American Physicians* 10 (1895): 57–59.

Stokes, Joseph Jr., John A. Farquahar, Miles E. Drake, Richard B. Capps, Charles S. Ward, and Albert W. Kitts. "Infectious Hepatitis: Length of Protection by Immune Serum Globulin." *Journal of the American Medical Association* 147, No. 8 (October 20, 1951): 714–719.

Stokes, Joseph Jr., Robert E. Weibel, Eugene B. Buynak, and Maurice R. Hilleman. "Live Attenuated Mumps Virus Vaccine." *Pediatrics* 39, No. 6 (December 1959): 363.

Strauss, John S., and Albert M. Kligman. "Effect of X-Rays on Sebaceous Glands of the Human Face." *Journal of Investigative Dermatology* 33, No. 6 (December 1959): 347.

Swain, Robert E., and W. G. Bateman. "The Toxicity of Thallium Salts." *Journal of Biological Chemistry* (December 9, 1909): 147.

Swyer, G. I. M. "An Evaluation of the Prophylactic Antenatal Use of Stilboestrol." *American Journal of Obstetrics* 58, No. 5 (1949): 994–1009.

Tate, Nick. "Clinton Calls High-level Conference on Radiation Testing." *Boston Herald,* December 31, 1993.

Tate, Nick. "Radiation Study Was Widespread at Fernald." *Boston Herald,* December 30, 1993.

Tate, Nick. "State Probes Report Patients Fed Radioactive Food." *Boston Herald,* December 28, 1993.

"The Law: Sterilized: Why?" *Time Magazine,* July 23, 1973.

Ticktin, Howard. "Hepatic Dysfunction and Jaundice in Patients." *New England Journal of Medicine* 267, No. 19 (1962): 964–968.

Triplet, Rodney G. "Henry Murray: The Making of a Psychologist?" *American Psychologist* 47, No. 2 (February 1992): 299–307.

Van Goozen, Stephanie, Walter Matthys, Peggy T. Cohen-Kettenis, Victor Wiegant, and Herman van Engeland. "Salivary Cortisol and Cardiovascular Activity During Stress in Oppositional-Defiant Disorder Boys and Normal Controls." *Biological Psychiatry* 43 (1998): 531.

Warfield, Louis M. "The Cutaneous Tuberculin Reaction." *Journal of the American Medical Association* 50, No. 9 (February 1908): 688.

Wells, Kathleen, and Lynn Sametz. "Involvement of Institutionalized Children in Social Science Research." *Journal of Clinical Child Psychology* 14, No. 3 (Fall 1985): 245–251.

Wilder, William H. "Report on the Committee for the Study of the Relation of Tuberculosis to Diseases of the Eye." *Journal of the American Medical Association* 55, No. 1 (July 2, 1910): 21

"Will Give Children Paralysis Vaccine." *New York Times,* August 21, 1934.

Williams, Jonathan H., and Walter Freeman. "Evaluation of Lobotomy with Special Reference to Children." *Association for Research in Nervous and Mental Disease* 31 (1953): 311–318.

Wilson, C. J. "Ready-to-Eat Cereals and Dental Caries in Children: A Three-Year Study." *Journal of Dental Research* 58, No. 9 (September 1979): 1853–1858.

Wong, Doris Sue. "DMR's Plan for Self-Probe Is Assailed." *Boston Globe,* December 30, 1993.

Woody, Kathleen. "Legal and Ethical Concepts Involved in Informed Consent to Human Research." *California Western Law Review* 18 (1981): 50–79.

## ARCHIVE COLLECTIONS

College of Physicians of Philadelphia, Philadelphia, Pennsylvania.

Documents on the history of Sonoma State Hospital at Bancroft Library, University of California, Berkeley, California.

Medical Archive, University of Pennsylvania Medical School.

National Archives, College Park, Maryland.

New York Academy of Medicine, New York, New York.

Peace Collection, Swarthmore College, Swarthmore, Pennsylvania.

Pennsylvania State Archive, Harrisburg, Pennsylvania.

Pennsylvania State Library, Harrisburg, Pennsylvania.

University of Pittsburgh Archives, Pittsburgh, Pennsylvania.

Papers of Henry Beecher, Countway Library, Harvard Medical School, Boston, Massachusetts.

Papers of Walter Freeman, George Washington University Archive, Washington, DC.

Papers of Saul Krugman, New York University Medical Archive, New York, New York.

Papers of Joseph Rauh, Library of Congress, Washington, DC.

Papers of Jonas Salk, Mandeville Special Collections Library, University of California, San Diego.

Papers of Joseph Stokes Jr., American Philosophical Society, Philadelphia, Pennsylvania.

## REPORTS

Advisory Committee on Human Radiation Experiments. *The Human Radiation Experiments. Final Report of the Presidents' Advisory Committee.* New York: Oxford University Press, 1996.

Presidential Commission for the Study of Bioethical Issues. *"Ethically Impossible" STD Research in Guatemala from 1946 to 1948.* Washington, DC, September 2011.

Report of Progress in Research. Report II, July 1, 1945 to June 30, 1946, Nutritional Biochemistry Laboratories, Department of Food Technology, Massachusetts Institute of Technology.

Task Force on Human Subject Research. *A Report on the Use of Radioactive Materials in Human Subject Research That Involved Residents of State-Operated Facilities within the Commonwealth of Massachusetts from 1943 through 1973.* Department of Mental Retardation, April 1994.

United States Congress, Senate Committee on Labor and Public Welfare. Subcommittee on Health. Quality of Health Care—Human Experimentation, Hearings, Ninety-third Congress, First Session, on S. 974. Washington, DC: U.S. Government Printing Office, 1973.

## ADDITIONAL SOURCES

Daly, Marie E. "History of the Walter E. Fernald Development Center," n.d.

Freeman, W., and James W. Watts. "Schizophrenia in Childhood: Its Modification by Prefrontal Lobotomy." Paper read at the Institute of Living, February 26, 1947, pp. 1–9.

Leung, Rebecca. "A Dark Chapter in Medical History," *60 Minutes*, CBS, February 11, 2009.

Nazi Doctors Trial. *United States v. Karl Brandt*, Nuremberg, Germany, June 13, 1947, pp. 9173–9174.

Tudor, Mary. "An Experimental Study of the Effect of Evaluative Labeling on Speech Fluency." Master of Arts thesis, Graduate College of the State University of Iowa, August 1939.

## INTERVIEWS

*Former Test Subjects*

Almeida, Joseph—September 7, 2009.

Anthony, Edward—November 8, 2012.

Chabasinski, Ted—March 15, 2010, October 26, 2011 and November 2, 2011.

Dyer, Charles—February 21, 2011 and November 2, 2011.

Frank, Leonard—October 4, 2011.

Larocque, Austin—September 11, 2009 and October 9, 2009.

Shattuck, Gordon—July 18, 2010, February 21, 2011 and March 9 & 25, 2011.

*Physicians & Scientists*

Ackerman, A. Bernard—February 1, 1996 and September 9, 1996.

Callaway, Enoch—October 25, 2011.

Gross, Paul—January 22, 1996.

Kagan, Avir—November 3 & 5, 2011.

Kelsey, Frances—January 12, 1995.

Ketchum, James—February 15, 1995, February 6, 1996 and October 25, 2011.

Kligman, Albert—January 10, 1995.

Koprowski, Hilary—September 14, 2009.

Lisook, Alan—January 13, 1995, March 23, 1995, and January 19, 1996.

Livingood, Clarence—September 4, 1996.

Maletskos, Constantine—February 22, 2011 and April 28, 2011.

Southam, Chester—July 28, 1996 and October 3, 1996.

Urbach, Frederick—February 7, 1996.

Wecht, Cyril—August 5, 2009.

Wood, Margaret Grey—February 2, 1996.

*Additional Interviews*

Alves, Karen—November 2, 2011.

Arbiblit, Don—July 1, 2009.

Bly, Jessie—May 6, 2010 and November 9, 2012.

Borseth, Eric—July 29, 2009.

Clapp, Pat—October 12, 2011 and August 5, 2012.

Cowen, June—August 28, 2009.

Hagerty, Dennis—October 30, 2011.

Hahn, Christine—August 7, 2011.

Kaye, Jeffrey—October 26, 2011.

Levin, Joseph—August 31, 2009 and May 3, 2010.

Lombardo, Paul—August 13, 2010.

Lurz, Paul—April 21 & 26, 2010.

Marlow, Sandra—April 10, 2010.

Milstein, Alan—July 1, 2009.

Misilo, Fred—August 28, 2009 and April 26, 2010.

Oaks, David—March 15, 2010.

Parascandola, John—June 6, 2010.

Power, Jack—November 27, 2012.

Scheberlein, Robert—May 3, 2010.

Scull, Andrew—August 15, 2011.

Sharav, Vera—September 21, 2011.

West, Doe—April 20, 2010 and April 28, 2011.

# INDEX